I give the warmest and most cordial welcome to the new book by our good colleague and friend, Vladimir Lugo: CYBERMATURITY. This work explores the theme of cyber maturity, which is not confined to the world of technology but breaks through with wisdom and relevance in the realms of personal, professional, and ministerial growth.

This unique book connects the cyber world with emotional, spiritual, and technological development. Lugo has the ability to explore and analyze the important subject of the virtues of technology in service of a relevant and mature ministry. According to our author, this necessary and appropriate type of growth is based on the assimilation of important theological values of personal and professional maturity, and understanding the ministerial challenges manifested in contemporary society.

I highly recommend reading this important book by my colleague Vladimir, as it provides an intelligent and relevant contribution to contemporary technological ministry, grounded in accurate and intelligent theological analysis. It is a book for ministers and laypeople alike, as well as for individuals at various levels of technological and spiritual development and maturity.

I congratulate Vladimir for sharing this magnificent work with us, which breaks through both the world of technology and theology.

Dr. Samuel Pagan
Hispanic Programs Dean
Center for Biblical Studies in Jerusalem
Jerusalem, Israel and Lakeland, Florida

Vladimir Lugo has done it!
He has given us a valuable tool that serves as a bridge between the 20th-century church and the 21st-century church. It's time for us to utilize everything within our reach to expand the Kingdom of God.

CYBERMATURITY is the first tool, hopefully one of many, that provides us with a clear, simple, and theologically aligned process to approach technology with maturity and integrity. This resource will transcend generations as the initiator of an ecclesiastical movement for a new evangelism and discipleship, using everyday technology to bring eternal value to those in great need.
Move forward, towards maturity!

Rev. Carlos A. Velez
International Speaker, Church Planter, and Senior Pastor
Christian Missionary Alliance
Charlotte, North Carolina

CYBERMATURITY, by Vladimir Lugo, is a book that tenaciously addresses the discussion about the lack of more conscious and intentional use of technology for the development of Christian ministries. The author combines his biblical-theological knowledge with extensive professional experience in media and information systems. One of the strengths of the text lies in pushing the reader to go a step further, to step out of the comfort zone and embark on a challenging and diligent growth adventure in the efficient management of computer resources and social networks. Each chapter guides the reader through practical and achievable steps to reach that goal. The text points towards a hopeful future for those churches and ministries that desire to advance towards a productive cybermaturity.

Valentin Gonzalez-Bohorquez
BIOLA University
Professor, Author, and Journalist
Author of over twenty books on history, literary essays, poetry and short stories.

In times like the ones we are living in, all institutions have immersed themselves in a sea of technology, often without knowledge or training, resulting in disappointing, if not frustrating, outcomes. The church is no exception. The desire to bring souls to Christ has led us to plunge into this sea of information without knowing which direction to row.

The book CYBERMATURITY by my friend Vladimir Lugo is a powerful, useful, and practical tool for every church, pastor, ministry leader, and production pastor who wants to fulfill their goals and meet the congregation's expectations.

With simple yet profound explanations, concrete and enlightening examples, I believe that this book, unique in its kind, comes to build a new culture of technological maturity that we never even dreamed of, not realizing how desperately we needed it.

Thank you, Vladimir, for contributing to the Great Commission in the field of technology like no one else could. I believe that the history of technology related to the church will have a before and after this book.

Hector L. Blancovitch
Executive Pastor
Catacomb 5 Christian Ministry
Añasco, Puerto Rico

With enthusiasm and a highlighter in hand, I dove into the task of reading this fascinating book. Quickly, I realized that within it I found an abundance of wisdom and experience: revealing phrases, irrefutable statements, biblical context, and powerful strategies to transform (once and for all!) many of the technological programs and projects that we continuously start (or abort!) in churches. Without a clear direction or objectives, these endeavors often become true pains in the neck for pastors, treasurers, and us technicians!

I have laughed and enjoyed the anecdotes, but I have also been seriously confronted with the truths it presents. Finally, someone has challenged me to mature! I have learned a great deal about the need to approach new projects differently, centered on people, with real objectives, proper planning, and measurable results.

The book delves into clear and current topics, skillfully supported by biblical foundations, written by an expert teacher who effortlessly navigates and delves into the worlds of technology, theology, and ministry work. I confess I have been tempted to use some of these topics as the basis for some of my sermons!

I believe that any church or organization about to embark on new technological projects or in need of reorganizing existing ones (e.g., websites, audiovisual production for TV, radio, and the internet, administrative and membership systems) should immediately make use of this valuable tool. It is time to give space to these voices, prepared by God, to urgently help us ensure that technology in the church serves the "fundamental purposes" of the ministry of the Word, "without losing focus"... "in order to bear fruit."

I highly recommend it!

Nelson Lainez
Sound Engineer
Fontana First Assambly of God
Fontana, California

Congratulations! It is a great piece of work and a good start for the experience of creating continously. As a reader, I both struggled with it and enjoyed it. There are terms with which I am not familiar, which challenges me, and that is a positive aspect of the book. As I progressed, I liked it evern more, especially when you apply biblical criteria amidst technological criteria. I do not know how people will digest it or if they will consider it applicable right now, but that is not a flaw, but rather a wake-up call for those of us involved in ministry to connect more with the reality that surrounds us.

Harold Caicedo
Senior Pastor and Minister of the Word
Christian Reformed Church of North America
Fontana, California

What impacted me the most about this book is how it breaks paradigms regarding what technology, innovation, perfection, and their implementation within ministry are. Additionally, the step-by-step approach and the detailed and finely explained concepts brought me closer to a greater understanding of technology hand in hand with maturity. It also provided valuable models and maps to intertwine technology as a means to an end and technology as a human activity.

This is a relevant book for all church leaders, and I want to highlight the depth of each theological explanation by Vladimir. This book challenged me as a professional in the field and as a Christian to delve deeper and bring these concepts to my local church in Brazil.

Franklin Vera Cardona
Systems Engineer
Connection Group Leader
Rio de Janeiro, Brazil

VLADIMIR LUGO

CYBERMATURITY

7 **STEPS** TO UNLOCK THE POWER OF TECHNOLOGY FOR FRUITFUL MINISTRY

SERIES: CYBERMINISTRY **2**

CYBERMATURITY

7 Steps to Unlock the Power of Technology for Fruitful Ministry

vladimir@vladimirlugo.com
www.vladimirlugo.com

First English Edition published in the United States of America
by Tepui Media, LLC, Sheridan, Wyoming

ISBN-13 Paperback: 978-0-9973369-8-6
ISBN-13 Electronic: 978-0-9973369-9-3
ISBN-13 Audio: 979-8-9884611-1-1

Library of Congress Control Number: 2023909759

Other resources by Vladimir Lugo

Cyberminister Today
Receive my articles where I explore the topic of using technology in ministry and receive notifications about new publications, conferences, and seminars right in the comfort of your email inbox. Visit www.vladimirlugo.com.

#ministerialtechnologies | #cybermaturity

Follow me **@vladimirlugomt** and join the conversation in social media.

DEDICATION

To the three woman
in my life.
I am not Peter Pan's envy
because of you.

Maturity begins to manifest itself when we feel that our worries about others are greater than our worries about ourselves.

- Albert Einstein

TABLE OF FIGURES

Figure 1: Comparing Maturity Models
Figure 2: Maturity Roadmap
Figure 3: Attitudes Scale
Figure 4: Attitudes Scale Example
Figure 5: Attitudes Scale Totals by Team
Figure 6: Attitudes Scales Totals by Team Example
Figure 7: Principles Lanes
Figure 8: Questionnaire: Principles Lanes of Technology Maturity
Figure 9: Principles Lanes Totals by Team
Figure 10: Principles Lanes Totals by Team Example
Figure 11: Strategic Areas Table
Figure 12: Technology Wheel
Figure 13: Technology Wheel Example
Figure 14: Technology Wheel Totals by Team
Figure 15: Technology Wheel Totals by Team Example
Figure 16: Reverse Engineering
Appendix A: Attitudes Scale
Appendix A: Attitudes Scale Totals by Team
Appendix B: Questionnaire: Principles Lanes
Appendix B: Questionnaire Key
Appendix B: Principles Lanes Totals by Team
Appendix C: Technology Wheel
Appendix C: Technology Wheel Totals by Team
Appendix E: Consultants Interview

TABLE OF CONTENT

Dedication **vii**

Table of figures **xvii**

Table of content **xix**

Preface **xxiii**

Why This Book? xxiv

Who Should Read It? xxv

How to Use It? xxvi

Introduction **3**

Maturity **9**

Be Fruitful 10

Be Spiritual 13

The Fruit of the Spirit 15

The Fruit of Technology 16

Immaturity **21**

Growth Alternatives 22

Immaturity Attitudes 24

Maturity Avatars 31

Technology **35**

Technology Familiarity 36

What is Technology? 38

Purpose of Technology 41

Technology in Ministry 45

Maturity Models 51

Maturity in Life 53

Maturity in Technology 54

Parallels 59

Maturity Principles 61

Maturity Roadmap 65

The Leader Follows the Map 66

Maturity Requirements 67

7 Steps for Fruitful Ministry 70

God's Maturity 87

The Most Perfect 89

The Ancient of Days 90

The Most Fruitful 92

The Most Mature 95

Cybermaturity 105

Storms in Ministry 106

Cybernauts 108

Cybernetics 109

Cybermaturity 114

Benefits 120

Assessing Maturity 121

Teamwork 124

Attitudes Scale 126

Principles Lanes 131

Technology Wheel 138

Recommendations 144

Applying Maturity **149**

Reverse Engineering 150

Digital Presence 152

Custom Roadmap **163**

I Am Lost 164

Calling a Friend 167

Consultancy Biblical Manual 168

Custom Maturity Roadmap™ 174

Appendix A: Attitudes Scale **179**

Scale: Attitudes towards technology maturity 179

Scale Totals 180

Appendix B: Questionnaire **181**

Questionnaire: Principles Lanes 181

Questionnaire Answer Key 183

Questionnaire Totals 184

Appendix C: Technology Wheel **185**

Assessment: Technology Wheel 185

Technology Wheel Totals 186

Appendix D: Discussion Questions **187**

Chapter 1: Maturity 187

Chapter 2: Immaturity 187

Chapter 3: Technology 188

Chapter 4: Maturity Models 189

Chapter 5: Maturity Roadmap 189

Chapter 6: God's Maturity 190

Chapter 7: Cybermaturity 191
Chapter 8: Assessing Maturity 191
Chapter 9: Applying Maturity 192
Chapter 10: Custom Roadmap 193

Appendix E: Consultants Interview 195

Acknowledgements 197

Consulting Services 199

About the Series 201

About the Author 203

Limitation of Liability 207

Bibliography 209

Books 209
Websites 211

Endnotes 213

PREFACE

The one who knows how to swim can retrieve pearls from the depths of the sea; the one who does not would drown. That is the reason why only those who have the proper training should take the risk.

– MAIMONIDES, JUDEO-ANDALUSIAN THEOLOGIAN.

The digital era digital is an era of great challenges and opportunities for the church. However, few Christian thinkers have taken on the task of helping us navigate the stormy waters of contemporary technology, leaving us adrift.

Most church leaders have been trained in the study of the Bible, theology, and pastoral care, but not in technology matters, and gradually they have fallen behind, risking becoming captains of the past, unable to offer a response to the world of today. This needs to change.

If you are a Christian leader and experience overwhelm, uncertainty, and fear when considering using technology in your ministry, this book is for you. Within its pages, you will acquire the compass and grow in knowledge to steer the boat of your ministry to retrieve the hidden pearls in the depths of ministerial technologies and back to the safe harbor of your vision. You will be able to understand technology at each of its seven stages of maturity while gaining the necessary tools to assess your own technological state and course-correct with a solid foundation of biblical and theological principles.

Why This Book?

Cybermaturity is the only existing manual on the maturity of technology in the pastoral and Christian ministry. It has no comparison or precedent with any other work on the subject. It fills a gap in Christian literature regarding the use of technology in ministry, establishing a new category of reference: ministerial technologies.

Cybermaturity is the second volume in the *Cyberministry* series, which studies ministerial technologies from various angles. Each book explores a different dimension of technology in mission. This book will not provide you with the technical details of implementation or technological development; for that, you will need the other books in the series. However, it will help you gain a panoramic view of technology in ministry.

The concepts in this book have matured from a series of conferences on the subject that I shared in 2015 during *Lidervision* and the National Youth Leadership Forum, hosted by *Editorial Vida* and *Especialidades Juveniles*, respectively.

Although there are no existing maturity models for the use of technology in ministry, simply placing these words together ("maturity" and "technology") in the same sentence provokes insight in the minds of the Christian leaders with whom I have shared this topic. It makes automatic sense. Mature ministerial technology is much more effective, creates a greater impact, and is easier to manage than immature technology. I hope this book also creates many connections in your way of thinking and leading the use of technology in your ministry.

Who Should Read It?

The book is aimed at Christian leaders and workers who need to take the helm of their technological initiatives and programs.

Pastors

Those who desire to build a solid and relevant ministry in the current times. Senior pastors, administrative pastors, and pastors of various ministries can share a common vocabulary and form an innovative force.

Denominational Directors

Those with the challenging task of leading pastors and establishing policies and procedures that guide their flock. Regional and national directors, leaders of ecclesiastical centers and pastoral associations, all can gain a strategic understanding of the scope of technology in ministry.

Christian Educators

Those involved in the theological and ministerial training of our leaders, shaping our way of thinking. Professors in theological seminaries and universities, teachers in academic schools and Bible studies, can all guide their students towards technological maturity.

Leaders

Those who carry out practical ministry tasks week after week. Worship leaders and musicians, sound and media production teams, evangelism coordinators, missionaries, counselors and social workers, small group facilitators and hosts, administrative and managerial staff, can find practical advice to be more effective in their ministerial work.

All Christians

Those of us who have received a model of maturity in the person of Jesus Christ and have been called to the Great Commission. Before maturity applies to technology, it applies to the individual. Many pieces of advice in this book have application not only to technology in ministry but also in various areas of life.

Cybermaturity is intended for all those in leadership positions in Christian organizations of diverse nature, whether in churches, missionary and para-church organizations, educational institutions, or community service foundations, where there is a desire to make the best use of technology in carrying out their functions.

How to Use It?

Cybermaturity is written with four dimensions in mind, aiming to help us think biblically, theologically, and theoretically, and to apply the concepts to the practice of ministerial technologies:

1. **A Biblical and Theological Dimension**, allowing members of your ministerial team to regain the necessary confidence to advance in the development of the mission, enabling you to distill principles and develop practices endorsed by the Word of God.

2. **A Theoretical Dimension**, helping you put technological maturity into perspective, considering existing concepts common in the information industry and creating a framework for thinking and learning to lead your technological projects.

3. **A Practical Dimension**, so that you can take the assessments and exercises contained in this book, along with the discussion questions, and apply them promptly to the way you already use technology.

4. **A Human Dimension**, because we often forget the role that people play in technology matters, and as I often say, technology is by the people and for the people. In this, technology aligns perfectly with the mission of the Church.

Cybermaturity consists of ten chapters that progressively build the concepts of technological maturity, starting from its definition and biblical foundation to considering the need for practical application in the specific context of your ministry. For this reason, I recommend reading them in the presented sequence.

You will then find five appendices containing three assessments to evaluate technological maturity in your ministry, a series of chapter-segmented questions to stimulate group discussion, and a guide for assessing those who wish to work in the technologies of your ministry as consultants.

I invite you to contact me personally. You will find my contact information in the **About the Author** section. You can also join the book discussion group (https://goodreads.com/vladimirlugomt) on *Goodreads* and follow me on various social media platforms using the unique identifier @vladimirlugomt.

Furthermore, I would love to hear your feedback on the book. During my years in seminary, I learned that every book is a conversation. So, this is my invitation for us to engage in this relevant and necessary topic. Please, after finishing your reading, leave your comment about the book on the Amazon page. Your honest appreciation and feedback can help me better serve you.

With great expectations, your servant,

Vladimir Lugo
Los Angeles, California
May 2023.

CYBERMATURITY

INTRODUCTION

Let all the children grow. All but one.

– J.M. BARRIE, PETER PAN.

I was so frustrated that day. My eldest daughter had entered adolescence and was starting to fight what I like to call her "Independence War." Every man and woman have fought this war. I called a friend, and although I did not call him to talk about this, it was almost inevitable. He listened to me patiently for several uninterrupted minutes. Then he gave me the following nugget of wisdom: "They will grow up; the question is whether or not you will."

To grow or not to grow, that is the dilemma. Scottish author Sir James M. Barrie had the peculiar opportunity to define the psychology of this generation. Unbeknownst to him, his children's play *Peter Pan*, the boy who did not want to grow up, became the drama enacted on the stage of a society that exalts youth as an ideal and evades the challenges of adult life.

Resisting maturity is charming as a fictional story. But it stops being funny when it comes to real people or organizations. No pastor would name their church Peter Pan Church. No Christian would proudly say, "I attend the Church of Imminent Immaturity."

This also applies to ministerial technologies. We are obligated to grow in our technology programs. Throughout my experience as a computer professional and a minister by vocation, I have seen many technology implementation programs in churches and Christian organizations that start with good intentions, only to end in the frustration of false expectations and unfulfilled promises.

Immature technology will waste your resources, cost more to maintain in the long run, or cause you to miss precious opportunities.

Mature technology, in contrast, will begin with the end in mind, make an upfront calculation of the investment, set realistic goals, identify the desired outcome, and know when it has been achieved. In other words, it will strike a balance between costs and results to determine if it is worth it and if it can be completed.

Jesus Christ expressed it this way:

> Suppose one of you wants to build a tower. Won't you first sit down and estimate the cost to see if you have enough money to complete it? For if you lay the foundation and are not able to finish it, everyone who sees it will ridicule you, saying, "This person began to build and wasn't able to finish" (Luke 14:28-30).

Immature technology is incomplete; it will leave you halfway and can make you a laughingstock. The worst part is that you will not even recognize that you are halfway there. With mature technology, you can bear much fruit—fruit that you can count and share for others to count and share.

Immature technology does not consider sustainability models or the impact on organizational culture. It only focuses on what it takes to go live. On the other hand, a mature technology considers technical support, maintenance, and training, which are vital for changes to take root and be reflected in the community.

For example, it is not enough to have a television program. If you use second-rate production, the low quality of the program will cost you viewership, your reach opportunity will fade away, costs will accumulate, and you will waste a lot of time. But if you have maturity in mind, the production and programming will be top-notch, gaining viewership, opening doors, attracting sponsorship, and having the influence you expected.

The great challenge is, therefore, to achieve technological maturity in ministry. For that, we need a maturity roadmap. Only with a proper maturity roadmap can we ensure that ministerial technologies yield results.

Although there are various models of technological maturity in the IT and business management industry, leaders of Christian organizations generally have little knowledge of the subject.

The most immediate response I receive when I mention the words "maturity" and "technology" together is one of astonishment and questioning at first, then insight. Take any seasoned (mature) individual in technology matters, and they will be familiar with the concept of technological maturity. Take one of our Christian leaders, and they will likely show confusion.

Only with a proper maturity roadmap can we ensure that ministerial technologies yield results.

As you can see in the model I present in the following pages, moving away from that ignorance is the first step towards maturity.

In this book, you will discover the need for maturity in your ministry's technology programs. You will also learn about a model of ministry technology maturity, possibly the first and only one to date, and you can use this model to practically evaluate your own technology initiatives and programs.

My goal is not to provide all the answers regarding the topic, but rather to offer a framework for the questions and provide a language and model to help you understand technology through a ministry lens.

If we want to bear much fruit with our ministry technology programs, we not only need to understand the need for maturity but also have a clear model of maturity that allows us to evaluate our current state and take actions to grow. No ministry technology program will grow on its own.

If we do not have a clear vision of the future we want to build, if we are not intentional in technological development, and if we do not plan the level of maturity we want to achieve, an uncertain destiny awaits us.

If we seek to avoid wasting our resources, if we do not want to lose hours of sleep, and if we desire our leaders, servants, and volunteers to accompany us on a long ministry journey, we need to mature.

That is why I invite you to mature.

MATURITY

Do not fear moving slowly;
only fear standing still.

– CHINESE PROVERB.

The person who understands evil within their own heart is the only person who is useful, fruitful, and trustworthy in their values and obedience.

– JOHN OWEN, ENGLISH THEOLOGIAN.

Moving from the house we had just sold was really tiring that day. I focused on the tasks of dismantling, loading the truck, and leaving everything in relative order for the new owners. Nothing out of the ordinary, but exhausting. If you have ever moved from one place to another, you know what I am talking about.

We were emptying the house where our two daughters grew up. It was a difficult decision because we had become attached to the place, but there was no turning back. We had made the decision to sell, emotionally detached ourselves, and looked for new horizons while planning our life in the new city.

As I gave a final inspection to every corner of the house, nostalgia suddenly overwhelmed me. The simplest thing you can imagine brought tears to my eyes, and I could not hold back.

A cascade of memories gathered and came out through my eyes when I saw the growth marks on the door frames of my daughter's respective rooms. The dates accompanying each little line constituted a historical record of how much they had grown. I wiped my tears. I took out my phone. And I took photos of the door frames.

My daughters can tell you. I often threatened to punish them for disobeying me when I forbade them from growing, but they did not listen to me. They continued in their quest to grow and became grown up virtuous women. Joking aside, I am grateful for that.

Maturity is the natural path of everything that has life. We cannot remain attached to the past while expecting to mature. To mature is to move forward. To mature is to progress.

We need to dry our tears, treasure the important moments of the past, and embark on a future of growth and maturity. But what is maturity?

Be Fruitful

Maturity is the state of maximum vitality.

Vitality is the capacity for maximum reproduction.

Reproduction is the quality of bearing fruit.

This implies that maturity is the state of maximum fruitfulness, of maximum results. It is the level at which one can yield much fruit.

The Bible encourages us to bear fruit. It is an intrinsic quality of God's creation. God has endowed living beings with the ability to bear fruit and reproduce.

Genesis 1:29 reads: "Then God said, 'I give you every seed-bearing plant on the face of the whole earth and every tree that has fruit with seed in it. They will be yours for food.'"

Genesis 2:9 also reads: "The Lord God made all kinds of trees grow out of the ground—trees that were pleasing to the eye and good for food."

Not only the fruit, but the first fruits were considered an offering to God. The firstborn of the womb, the first fruits of the field, and the initial harvest were dedicated to God. The dedication of the first fruits was an offering of maturity.

By the way, God wants us to give Him the first fruits so that we may understand His heart, for He also gave us His first fruits when He gave us His firstborn.

Bearing fruit is the first commandment.

Perhaps it is not the first commandment in terms of priority or importance, but it is the first commandment chronologically. God said to man:

Genesis 1:28: "Be fruitful and increase in number; fill the earth and subdue it."

It was also the first commandment that God gave to Noah after the flood when He blessed him and his sons with similar words:

Genesis 9:1: "Be fruitful and increase in number and fill the earth."

Maturity is the state of maximum vitality.

Multiplying our businesses, our relational influence, the impact of our message of faith, our finances, and the reach of our ministries are part of the first commandment.

We do not reach our maximum capacity for reproduction until we mature. This is the reason for the call to maturity:

A mature person is the most fruitful person.

Maturity should permeate and be reflected in all areas of life and ministry because maturity leads to fruitfulness. The Lord Jesus said:

Matthew 7:16a: "By their fruit you will recognize them."

A mature person is recognized by their fruits. A mature ministry is recognized by its fruits. A mature plan for ministerial technologies is recognized by its fruits.

Maturity, and not just numerical growth, is the most important proof of the vitality of a ministry.

A mango tree can be mistaken for another tree, but not when it is loaded with mangoes. A mango tree reaches maturity at eight years old and produces up to 7,000 mangoes for the next 10 to 15 years.

A tomato seed can easily be confused with other legume seeds, but not when it is planted, matures, and bears tomatoes. The seed does not produce tomatoes because it is not mature, but it does produce tomatoes when it becomes an adult plant, and it produces a lot of tomatoes when it reaches its highest level of maturity.

The same can be said of your technology programs in ministry. Isn't it the purpose of every Christian enterprise to bear abundant fruit? An early-stage technology program in ministry may not show many results at first, like a small seed, but as it matures, it begins to bear some fruit, and it bears much more fruit as it reaches higher levels of maturity.

Be Spiritual

Character maturity is the result of the work of the Holy Spirit.

Galatians 5:22-23: "But the fruit of the Spirit is love, joy, peace, forbearance, kindness, goodness, faithfulness, gentleness, and self-control. Against such things there is no law."

A mature person is the most fruitful person.

The fruit of maturity is not just about having nice feelings. We often think that love, for example, is a commendable feeling. Nothing could be further from the truth. Love is action. The same can be said for each of the characteristics of the fruit of the Spirit.

Maturity is translated into good deeds.

Titus 2:7: "In everything set them an example by doing what is good."

Therefore, if the maturity of character is the work of the Holy Spirit, we can say that the maturity of the church is also the work of the Holy Spirit, the maturity of ministry is also the work of the Holy Spirit, and the maturity of technology in ministry is also the work of the Holy Spirit.

Often, I find that there is a disconnect between what is considered spiritual and what is considered worldly. We think that spiritual matters are related to praying, worshiping, singing, fasting, and reading the Bible, while working, going to school, personal care, or using money are worldly things. This way of thinking has caused the church to withdraw from certain areas and has resulted in a detriment to society.

Working, going to school, and wisely using our finances are spiritual matters. Exercising and eating healthily are spiritual matters because they take care of the temple of the Spirit. Having balance in life is very spiritual. Using technology in a holy manner for ministry purposes is also very spiritual.

The key lies in the fruit. Being spiritual is being mature. A mature person is the most spiritual person.

It is appropriate to think that your technological development will foster, strengthen, and affirm your spirituality and that of your organization. As you take new steps in your journey towards technological maturity, you also reach new levels of spirituality.

Of course, this does not happen by chance. Not all technology or all uses of technology lead to spiritual growth by inertia. You need to be intentional and have clear spiritual purposes for your technological development.

In other words, your technology programs should be planned, coordinated, and carried out with the aim of achieving spiritual advancements. Not because of technology itself or because other ministries are using it, but because of the specific spiritual objectives that we want to reach.

A mature person is the most spiritual person.

Your technology programs need to consider Paul's words to the Galatians and align themselves with the fruit of the Spirit, bringing forth love, joy, peace, patience, faithfulness, goodness, gentleness, and self-control.

Let us consider this idea a bit further.

The Fruit of the Spirit

Most people I share about the fruit of the Spirit think of it either metaphorically or in terms of how enjoyable it is to partake of its benefits.

If you ask a hungry person, they will be happy eating some fruit —it is nutritious, it satisfies the hunger, and it is sweet to the taste. This is an experience common to all. This is why it is easy to relate the fruit of the Spirit based on the benefits it can bring us. And there is nothing intrinsically wrong with this understanding, except that it falls short.

Certainly, if we love more, we experience more joy, we have more peace, we exhibit more patience, we face life's difficulties with greater faith, people are kind to us, everyone is good, exhibit gentleness, and exercise self-control, we will live a better life. It is like enjoying a sweet and desirable fruit.

But the true power of the fruit, though it is good nourishment (and we should have five servings a day), lies not in its nutritional value but in its reproductive ability, its reproducibility.

The fruit's natural purpose is to protect the seed. I cannot think of a way to shout this on paper, so I am going to say it again:

The fruit's natural purpose is to protect the seed.

The fruit is a delicious shell to protect the reproductive capabilities and preserve the genetic load that will be transferred to the next generation.

This is truly powerful. You can eat an apple and discard the core, as we often do. But within that core, there are potentially thousands more apples.

If you ask a mature person, a fruit is good for extracting the seeds, planting them, cultivating them, and producing much more fruit. The man at the supermarket buys the fruit to eat it, but the woman farmer buys the fruit to plant it and produce a hundredfold.

The Lord understood this well when He emphasized the importance of the seed in the parable of the sower:

Mark 4:20: "Others, like seed sown on good soil, hear the word, accept it, and produce a crop—some thirty, some sixty, some a hundred times what was sown."

The fruit of the Spirit is not only for your enjoyment but for it to reproduce fruit in others with the same genetic load.

The same can be said of technology. The fruit of technology is not only for your enjoyment but for it to reproduce fruit with the same genetic load in others.

The Fruit of Technology

The concept of the fruit of the Spirit has been a part of my Christian journey from an early stage. It is a concept that continually inspires and stimulates me. However, it was only recently that I thought of combining this concept with that of technological maturity.

What if we apply the concept of fruit to ministry technology? We would have more mature programs that fulfill more concrete and elevated spiritual purposes. Let us explore further.

Love

The use of technology in ministry should be motivated by love for God and love for others. It should produce love within us and replicate love in others. Love is necessary to reach people where they are, to continue loving amidst trials and challenges, and to cover the many mistakes made along the way. As you can see, love is a recurring theme in other sections of this book.

Joy

The use of technology in ministry should bring joy to us and replicate joy in others. Sometimes, using technology can cause frustrations, bitterness, and despair. Yes, it happens to Christians too. However, if we keep the reproductive capacity in mind –the results, the fruits– it will place joy at the center of our experience. It is truly satisfying when technology functions as designed and fulfills the purposes for which we use it.

Peace

The use of technology in ministry should bring peace to us and replicate peace in others. Peace will help you overcome the anxiety of keeping up with the latest trends and the worries of staying up to date with technological advancements. Peace will give you the assurance of reaching your destination if you know where you are going. The maturity roadmap can help you define your technological destination.

Forbearance

The use of technology in ministry should cultivate patience in us and promote patience in others. Patience is crucial when it comes to technology, as it requires time to learn, go through trial and error, and overcome obstacles before we can enjoy the joy of the results. Patience allows us to persevere and navigate through challenges with a steady and determined mindset.

Faith

The use of technology in ministry should cultivate faith in us and inspire faith in others. Faith is an active force that plants a seed today, trusting in the expectation of future results. Nothing is closer to faith than maturity, including technological maturity. The path to technological maturity does not assume immediate gratification but rather deferred gratification, combining it with patience and perseverance.

Goodness

The use of technology in ministry should cultivate kindness in us and inspire goodness in others. There are inherent dangers in the use of any technology that we need to be mindful of, to avoid promoting them and falling into them. Technology programs in ministry should be designed to do good without partiality or favoritism.

Kindness

The use of technology in ministry should cultivate kindness in us and promote kindness in others. Kindness differs from goodness in that kindness is an intrinsic characteristic of goodness, while goodness pertains to the external actions of goodness. Kindness is related to compassion or the ability to understand others in their condition of need and provide care. Technology in ministry should embrace kindness in its ability to know and understand others, an aspect that few ministries have fully explored using technology.

Meekness

The use of technology in ministry should cultivate meekness in us and promote meekness in others. Here is my definition of meekness: being teachable. A meek person acknowledges that they do not know it all and that there is much to learn. Technology can assist us in both aspects due to its ever-changing nature that demands staying updated and its capacity as an educational tool.

Self-control

The use of technology in ministry should cultivate self-control in us and promote self-control in others. Self-control is necessary in our interaction with technology. It is true that at times it can be frustrating when technology does not meet our expectations, but resorting to anger or losing control does not help. Technological processes rely on precision and control, just as maturity is the outcome of careful planning and guidance.

Ministerial technologies provide opportunities to exercise the fruit of the Spirit and to think creatively about how our interactions with technology can reproduce it in our hearts.

In conclusion, we recognize maturity by our fruit, and we recognize immaturity by our attitudes. As we need to understand the evil in our own hearts to be more fruitful, let us then discuss immaturity.

IMMATURITY

Maturity is to know when to be immature.

– RANDALL HALL, AMERICAN GUITARIST.

With immaturity comes uncertainty.

– WILLIAM L. NOLTE, AMERICAN SCREENWRITER.

My eldest daughter was two when she discovered wall art. She became a desperate street gangster decorating all the walls around the house with her multicolored graffiti. Every white open wall was an invitation, shouting to her: "Come paint me!"

I was close to changing jobs and becoming a painter. Yes, a painter of my own house, because she was quicker with the permanent marker than I was with the big brush.

During those days, we had just started to apply a new technique based on Thomas W. Phelan's book, *1-2-3 Magic.* This technique involves modifying behavior by giving the child three warnings and then a consequence.

My daughter took her set of markers and started drawing. My wife then repeated the familiar phrase: "We don't paint on the walls, stop painting there." My daughter glanced at her and, like any good artist, ignored her. Then my wife raised her voice: "Stop painting... One... Two... Three..." and before she could punish her by making her sit in a chair for a couple of minutes, my daughter responded half-heartedly: "Four, five, six, seven..." Total failure!

If you are a parent, you know how challenging it is to change our children's behavior with words alone. But we did not want to frustrate the childhood of someone who could potentially become the next Mary Cassatt. That is when I came up with an idea that would change everything.

I went to the hardware store, bought green chalkboard paint, and colored chalks. I chose a suitable wall and painted a large and accessible section. I took away my daughter's markers and handed her the colored chalks, saying, "From now on, you can draw as much as you want, but stay within the green area." It worked like magic.

Growth Alternatives

While preparing the talk that originated to this book, I asked my brother, "What is maturity?" and he jokingly replied, "Maturity is the absence of immaturity."

I then responded, "How immature of you!"

Over time, I have reflected on my brother's response and found that it holds some hidden wisdom.

For many people, it is difficult to recognize maturity. It is easier to identify immaturity when we see it. It is more obvious. Their graffities fill the walls with their contaminating presence.

In identifying the characteristics of immaturity in this chapter, I do not intend to focus on immaturity as a harmful way of criticizing the deficiencies or weaknesses of our leaders or ministries. Immaturity in our culture carries negative connotations. However, immaturity in a certain area can be the result of intentional decisions. When that is the case, it is not inherently bad.

Being able to identify the undesirable characteristics of immaturity helps us make a conscious effort to promote growth by fostering the behaviors we do want to see. It is more about setting clear boundaries against immaturity and providing alternative paths towards maturity. These growth alternatives are desirable replacement behaviors.

In other words, you do not strive to stop being immature, but rather to be mature, and as you grow in maturity, you naturally cast off immaturity.

The immaturity of our ministerial technology programs is a real problem.

The immaturity of our ministerial technology programs is a real problem. Perhaps we have not seen the need to mature in our technological initiatives due to certain attitudes that, since we are immersed in them, we do not recognize as immature. After all, what teenager likes to be told they are immature?

The bad news is that many ministerial technology programs (and well, it is worth mentioning that also many ministries, many churches, and even many Christians) remain immature.

The problem with immaturity is that it breeds uncertainty. It is very difficult to lead a ministry with uncertainty. Uncertainty lacks commitment and is abundant in risks.

Maturity, on the other hand, is inversely proportional to risks. The higher the technological maturity, the lower the risk of your programs failing. The good news is that immaturity can be corrected with growth alternatives.

Immaturity Attitudes

Here are some attitudes of immaturity that we need to overcome:

Fear

Graffiti: We may feel fear of failure, rejection, and criticism, and therefore, seek refuge in what is familiar or what we can control. Fear can originate from the sensation of having to lose control. At times, an inferiority complex can make us become suspicious and distrustful individuals.

Many reasons for feeling fear towards technology may be legitimate, but without proper investigation, they can be unfounded. The most common reasons I have heard regarding this matter are:

- "Technology is too expensive." This is one of the most frequent excuses, although there are many alternatives to minimize the economic impact of technology, such as using free software, second-hand devices, or good budget planning, for example.
- "No one knows how to use technology." This reason is easier to overcome. You can learn or hire someone who has the necessary knowledge and skills. When we hire someone, we accelerate the time to see results faster.
- "No one knows what this technology is for." As I will mention in **Chapter 4**, exposing oneself to knowledge is the beginning of the maturity process. This is one of the advantages of seeing maturity as a map that takes you from your present to your technological future.
- "Certain technology is against the Bible." The corollary to this statement would be "if you use it, you are sinning." There

have been no Bible studies from a technological perspective, which is the subject of other books in the series. On the contrary, the Bible has a favorable attitude towards technology. That is why this book contains many biblical references. Other myths about technology fall into this category.

- "It is difficult to obtain the technology." This is the issue of access. In some parts of the world, this is a reality that we cannot ignore. The impact of digital inequality on the economy and less privileged social groups is discussed. The concept of technological maturity is even more necessary in these cases.
- "The technology is not available in my language." Another real issue to consider. A technology maturity program can include multilingual services in its more advanced stages.

If you feel like you do not have the skills to embark on a technology program or that, if you start something new, people will criticize or judge you, let me tell you that you are not alone. We have all been there. If you feel like technology is a vast, unknown, and inhospitable territory, and it could be, and that as you walk through it, you will not be able to control every detail, it's true. At least in the beginning. But do not let fear paralyze and disable you.

Green Chalkboard: Some people think that hatred is the opposite of love. According to the Bible, the opposite of love is fear (1 John 4:17-18). So, the antidote to fear is love. We need to review our vision, mission, and working style in the light of God's love and love for our neighbors (Matthew 22:36-40). If we love our neighbors enough, we will do whatever it takes to demonstrate that love, even if it means using technologies that make us fearful, and then love drives out fear. Love leads us to find purpose because love is the engine of the mission.

The love of God motivates you to elevate the excellence of worship services in terms of music, instruments, sound, and sermons. Love for the people in your community compels you to maintain an updated and attractive website so they know where to turn in times of despair.

It is normal to feel some fear when experimenting with new technologies, but in those moments, we need to elevate our perspective and act in love, love for our neighbors in the pursuit of our mission.

Resistance

Graffiti: Some of the attitudes of immaturity are passive. Resistance, however, is active. It consists of putting obstacles in the way of technological initiatives and can manifest itself in various ways through delaying tactics and hindrances, such as boycotts, absenteeism, gossip, threats, hidden agendas, and even intimidation of leaders.

Whenever I am invited to give lectures on the topic, I like to ask: What are some obstacles that prevent us from implementing technology in ministry? The list of answers is not very long. Almost everything boils down to costs, lack of knowledge, or lack of trained personnel. You can see that each of these things can be overcome. Free technology can be used, learning can take place, and skilled people can be found or trained. The attitude of resistance will always find excuses for any proposal.

Green Chalkboard: I am inclined to think that if we are employing resistance to change tactics, it is because we do not desire to grow. Some of these attitudes are strongly condemned in the Bible and can even be considered sinful. Jealousy, strife, and divisions are on the same level as orgies, heresies, and witchcraft (Galatians 5:19-21). For that, we need to repent.

We need to become meek individuals. Being meek means being teachable. When we allow ourselves to be taught, we lower our defenses and abandon self-interest in favor of the mission. Then we obtain the resources to finance any technological initiative, learn the technology we need to learn, and connect with competent individuals who can guide us in our development.

Beign meek means being teachable.

Conformity

Graffiti: Conformity is related to maintaining the *status quo*, position, or established methods. It also involves being content with stability because we generally like change, but we do not like to change. The problem with this stance is that the world around us is not static and controlled but is constantly transforming, and if we do not grow, we will inevitably lose our current standing. Without maturity, the loss of the *status quo* is inevitable.

There is no clearer example of this dynamic than technology, and I will illustrate it with a simple example. You go to a store and buy a computer with cutting-edge specifications: the fastest and most efficient processors, maximum memory capacity, high-performance drives, and the most advanced operating system. After a few months, industry changes and even updates to the operating system make the computer outdated and slower.

Green Chalkboard: We need to review our theology of change and our ministerial philosophy to become entrepreneurial individuals who chart new courses. We need the adventurous spirit of an explorer, who is willing to try new methods to survive and reach new horizons.

The entrepreneurial church creates a balance between growth in numbers and character, between quantity and quality. It is not one at the expense of the other, but both working together. This means that ministerial technology multiplies the number of members, increases

Sunday attendance, and improves the quality of the services offered. The entrepreneurial church manages its resources to move towards new states of maturity.

Without maturity, the loss of the status quo *is inevitable.*

Blindness

Graffiti: Blindness has multiple dimensions. One may not see the true condition of the ministry, the prevailing needs of the community, or the changes in culture or demographic composition. In other words, the person is not interested in discerning the times. It is also related to a lack of a clear vision of what we want to achieve.

Blindness can be total or partial. We either fail to see the complete picture or can only see the next step. The blindness of some ministries in technology matters prompted me to write this book with the intention of providing a comprehensive, step-by-step roadmap of development and maturity. Of course, this roadmap will need to find specific expressions in particular contexts, but it remains useful in its general principles.

Green Chalkboard: Blindness is cured when our vision is restored. Vision is the fundamental element of every strategy, including technology strategy. We need new eyes, but how do we acquire new eyes?

We cannot discount the role of prayer. God can give us a spirit of wisdom and illuminate the eyes of our hearts (Ephesians 1:16-18). A retreat, for example, involving group prayer and meditation, can be the key that opens the door to vision in some cases. However, it is unlikely that a parchment with your technology plan will descend from the sky. Your vision and technology plan will require extensive research and teamwork.

Your team will need to conduct a multifaceted evaluation that includes assessing your current state, your place in the community, your cultural context, and the feasibility of certain technologies to discern the times in which you minister.

This opens the opportunity for new eyes to be brought in, meaning those of someone external to your ministry who can provide a neutral evaluation and offer specific recommendations. It is important to surround yourself with good and diverse advisors. A ministry technology consultant, as we will see in **Chapter 10**, a city official, or a demographic research agency can assist you in different areas and at different times.

The Bible provides us with a detailed list of all the men who gathered with David to hand over the kingdom of Saul. This list refers to the men from Issachar as "men who understood the times and knew what Israel should do" (1 Chronicles 12:32). Remember that the name "Issachar" means "rewarded." Being knowledgeable about the times and being able to guide decision-making brings many rewards.

Comfort

Graffiti: Comfort with what exists or what we already know how to do leads to stagnation. It can also generate an excessive attachment to tradition. Many follow the saying, "Better deal with the devil you know than the demon you don't." This saying contradicts a fundamental biblical principle, as we will see later: God is always doing something new, and God's new is always better.

It is normal for people to cling to the technology they are familiar with, but when excessive attachment to the familiar leads to ignoring the new, it becomes a limiting factor. Excessive attachment to outdated methods and programs, even when it has been proven that they no longer work, demonstrates immaturity. When you read **Chapter 5**, you will realize that ignoring and innovating are at the two extreme points on the maturity roadmap.

Green Chalkboard: What is now tradition was once considered innovation. We need to correct our theology to understand the role of innovation. That is what I am trying to do in **Chapter 6**.

> *What is now tradition was once considered innovation.*

It is essential to strike a balance between the fundamental principles of faith and the methods for sharing it. There are aspects of our essence that should not change, but there are others that need to be adjusted to remain relevant in a changing world. The principles are enduring, but the methods can change.

Sterility

Graffiti: Sterility refers to the inability to bear fruit. This occurs when our technology programs fail to produce good results or translate into good works. Often, this happens because we have not clearly defined what those good results or works should be.

Technology can serve many purposes in various contexts, but you need to define its purpose in your specific situation. Here are some ideas:

1. You want to improve the productivity of your team.
2. You need better coordination of activities and calendars.
3. You want to encourage effective and regular communication.
4. You aim to train others in ministry tasks.

5. You need to have a detailed understanding of your financials.
6. You are eager to increase donations and want to do it online.

The list is only limited by your imagination, but in each case, you can draw a direct line between the fruit you want to achieve and the technologies that can help you reach it.

Green Chalkboard: So, we need to define how success will look like when we achieve it. For each initiative, determine clear objectives, their ministerial purpose, and the technologies that will help you get there. For example, define how many people you want to train, for how long, using which educational technologies, and for what specific purposes in the ministry. Specify how much you want to increase online donation income, how you will adjust the budget, and what systems you will use.

Embark on a strategic planning process, which will help you establish clear goals and create an action plan to achieve them. You can read more about strategic alignment in Aubrey Malphurs' book on the topic, my own book *Cyberministry* (check the bibliography), and my blog article: *5 components your strategic plan cannot leave out.* [1]

Maturity Avatars

The word *avatar* has become popular in technology and computer science as the digital representation of a person. Avatars allow individuals and institutions to create a unique cyber identity that, in some cases, represents the real person, and in other cases, allows them to assume fictional personalities.

It is advisable to have a consistent digital personality, for example, on social media and websites. There are specialized software applications that can help you create and maintain your cyber identity or digital alter ego, such as the *WordPress* identity known as *Gravatar.* (https://en.gravatar.com).

The word *avatar* originates from Sanskrit and means "change" or "transformation." In Hindu religion, it refers to the incarnation of a deity. Please understand that I am not endorsing or promoting Hinduism with what I am about to say. What appeals to me about this concept is the ability we have to "embody" someone that we are not, yet we can become. This is what maturity is all about.

We can now adopt the attitude and mental disposition of a person determined to mature, an avatar of maturity. Fluidity and orientation will become posture and culmination. In other words, we can become what we desire. We are children, but we can be adults.

You might be thinking, "You are supposed to be talking about technology, why are you talking about personal attitudes?" If you are thinking like this, you are right. I need to give you an explanation.

On one hand, the concept of maturity is borrowed. It pertains to living beings, not inanimate objects. It is easy to understand maturity when applied to a person or fruit. A mature person is worthy of our trust. A ripe fruit is worthy of our palate.

On the other hand, technology does not exist in an inanimate vacuum. Technology is by the people and for the people. Technology thrives on the lifeblood of those who create and use it.

One of the most common mistakes I see in Christian organizations as a ministry technology advisor is their disregard for people.

We need to think about the people who will benefit from technology; those who will hear the message of the Gospel or receive discipleship training, for example. Caring for them would involve having clear ways to deliver the message, considering their ages or life stages when designing our communication. Caring for them would mean defining the content of discipleship and meeting with them every week to discuss it. All of this is mediated by technology.

We also need to keep in mind those who will be responsible for adopting, implementing, and maintaining the technology. The people who sustain ministry technology initiatives in the medium and long term are key to technological maturity. Caring for them begins with selecting qualified technical staff. Every technology requires expertise and the right people to maintain it. Caring for them means providing them with opportunities to stay updated with technology and explore creative ways to do ministry with it.

If we address the six attitudes of immaturity (fear, resistance, conformity, blindness, comfort, and sterility) in the people leading and operating technology in ministry, we will correct the course of technological maturity. The degree of technological maturity is directly proportional to the technological maturity of the people who support it.

When speaking of the degree of maturity, I specifically refer to the level of maturity achieved by following the roadmap in **Chapter 5**, which I apply to a concrete example in **Chapter 9**.

Maturity involves vision for the future, drive for the novel, exploration of the unknown, numerical growth, improvement in quality, humility to change, and meekness to be taught.

The degree of technological maturity is directly proportional to the technological maturity of the people who support it.

The Lord Jesus Christ, out of all the things He could have said for us to learn from Him, only said:

> Learn from me, for I am gentle and humble in heart (Matthew 11:29).

The less common but better translation of the word "gentle" here would be "meekness." Humility helps you recognize what level of maturity you are at. Meekness helps you rise to the next level.

Maturing is biblical, and not only biblical but also necessary. According to the apostle Paul, maturity allows us to have a clear vision of our progress and attain perfect knowledge:

> When I was a child, I talked like a child, I thought like a child, I reasoned like a child. When I became a man, I put the ways of childhood behind me. For now we see only a reflection as in a mirror; then we shall see face to face [VISION]. Now I know in part; then I shall know fully, even as I am fully known [KNOWLEDGE] (1 Corinthians 13:11-12).

Vision and knowledge are the results of maturity. If you want to have a clear vision and authoritative knowledge of the use and impact of technology in your ministry, you need to have a maturity model. You need to move from childhood to adulthood in matters of technology in ministry.

Before reviewing some existing maturity models, let us explore the topic of technology in the next chapter.

TECHNOLOGY

Technology is everything that was invented after you were born.

– ALAN KAY, AMERICAN JOURNALIST.

In 1972, Nolan Bushnell forever changed the landscape of children's play, with the advent of the first Atari video game console. Its first game, *Pong*, achieved global popularity in just three years and guaranteed the commercial success of a new industry.

Pong simulated a table tennis ball bouncing back and forth on a television screen, intercepted by two lines acting as paddles. The lines were controlled by turning the knobs on a console connected to the TV. It was easy to connect and start playing.

Other toys of the time took a back seat. Kites, marbles, yo-yos, and jacks were replaced by a successive race to improve graphics quality, interaction complexity, and the speed and capacity of consoles.

Today's video games are far from this initial simple version. Modern consoles are high-performance computers capable of "seeing" movements or tracking them via radio frequencies to "interpret" the player's intentions. Additionally, video games have proliferated and reached every electronic device, from mobile phones to smartwatches.

There is no doubt that video gaming technology has matured. Some games have transcended the purpose of entertainment and now stimulate physical agility, mental activity, or learning capacity.

Technology Familiarity

Technology is one of those concepts that is so familiar that it becomes difficult to define. This is because all technology progresses towards familiarity on its path to maturity.

We know that video games are technology, but we do not think of kites, marbles, yo-yos, or jacks as such. Take a moment to reflect on this. These traditional toys are also technology.

As children, we used to make our own kites. Ingenuity was part of the fun. We went through endless trial and error to discover the perfect combination of bamboo, paper, and glue. We spent hours determining the direction and strength of the wind to make it soar. And then we learned to perform tricks in the air. But long before our experiments, someone came up with the idea of modeling the first kite.

The same can be said for other toys. To play with marbles, we had to tame glass with fire. To play with yo-yos, we shaped wood on a lathe. To play with jacks, we made a ball out of rubber and poured a metal alloy into a mold.

The same analysis can be extended to other categories. No one thinks of a chair as technology, yet there was a time when chairs did not exist, and someone had to figure out how to balance them, both in the literal sense of how many legs are needed to provide the best support, and in the economic sense of production, determining the minimum number of legs to keep costs down.

Over time, a chair remains just a chair. However, an electronically programmable sofa with automatic massage and zero gravity position, now that is technology worthy of occupying the floor of *Brookstone*.

> *All technology progresses towards familiarity on its path to maturity.*

The same applies to the ministry of the church. There was a time when the church did not use musical instruments, for example. Now we associate pipe organs with traditional churches, but they were not always present. It took almost a thousand years for the first medieval churches to adopt the organ as part of their worship.

The organ is a tangible piece, and it is easier to see technology as something tangible, but the same applies to intangible techniques.

Today's evangelical church cannot conceive a worship service or evangelistic campaign without ending with an altar call. However, the altar call was introduced to the church by the American evangelist Charles Finney, who ministered between 1821 and 1875. The Christian church survived for over 1800 years without this revolutionary technique for its time.

Techniques are technology. Although these words share the same root and etymology, we do not think of them as technology. In this sense, ecclesiology, systematic theology, exegesis, hermeneutics, and homiletics that adorn our seminaries and prepare our pastors for ministry are technology, even though we may not think of them as such.

Therefore, we take for granted technologies that we grew up with and do not identify them as technologies, even though they are. This familiarity necessitates a clear definition of technology, or at least a definition around which we can create a consensus for ministry work.

What is Technology?

The definition we adopt for technology will inform and serve as a framework for our selection, development, implementation, and evaluation. It will also help us create a map of technological maturity that is relevant to the ministry context.

I opt for a pragmatic concept of technology. You can read about how I arrive at this concept in my book *Cyberministry*, and you can read more about the concept of technology in my article: *5 columns that sustain the use of technology in the Church.*[2] I will simply state it now:

> Technology is a cultural interconnected network of people, systems, processes, policies, and devices that allow us to interact with our surrounding world.[3]

This definition confirms that if something is systematized or standardized, it is technology. If a methodology is created for that something, it is technology. If it requires documented regulations and procedures, it is technology. And if it somehow involves interaction with the material world around us, it is also technology. These things are typically combined to create a finished product for mass consumption, but even if it does not reach the masses, it would still be considered technology.

Technology is an integrated ecosystem. People often rush to think that technology is only about devices, but devices have no context without policies, they do not yield results without processes, they do not interact without systems, and no one benefits if there are no people involved.

I want to illustrate this with two examples.

In 1996, the United States Congress passed a healthcare reform law to protect individuals. The law had two goals in mind. First, to prevent employees from being left without health insurance, and second, to protect the security and confidentiality of personal information and patients' medical history. This law is known as the *Health Insurance Portability and Accountability Act* or HIPAA.

At first glance, this may seem like just a law. But in the technology ecosystem, it is expected that this change in policies will impact the other four areas. And that is exactly what has happened.

Technology is an integrated ecosystem.

- In terms of people, HIPAA has required the training of specialists knowledgeable about the law to ensure that medical practice complies with it and remains protected by it. Every person with access to patient information is now responsible for understanding the law and safeguarding such information.

- In terms of systems, all systems, whether manual or automated, paper-based, or electronic, have had to adapt to comply with the law, creating a parallel technology industry. Security in accessing and handling information became a priority. Work methodologies changed.

- In terms of processes, every process in insurance companies and healthcare providers came under government scrutiny. Regular auditing mechanisms were established to ensure patient continuity and privacy.

- In terms of devices, computers had to implement stringent security measures, encrypt communications and data storage.

> Much software had to be redesigned to comply with the law, and industries were created for the secure disposal of information, among other things.

Each entry point in the technology ecosystem has repercussions on the remaining four areas. The same applies to the ministry of the church. Here is my second example.

How does a church decide that its worship service will be on Sunday at 10:00 am? I do not know. It could be due to tradition, consensus, or a decision informed by detailed research. It does not matter. What matters is that the worship service is scheduled at a specific time. We may not think of this as a policy, but it is.

The Merriam Webster dictionary defines policies as the prudence or wisdom in the management of the affairs.[4] This applies the same for an individual or an organization. From the decision about the schedule, the rest of the activities need to align.

- In terms of people, the pastor needs to prepare their sermon before Sunday, the musicians and singers need to rehearse the worship songs, the deacons need to purchase communion elements, and the ushers need to know the details of collecting offerings.

- In terms of systems, the pastor may use software for Bible study and the internet to read current topics they plan to bring on Sunday; the musicians and singers coordinate the use of sound and audiovisual systems, access sheet music, and follow a planning and direction method; the deacons need the purchasing card and know how to report expenses; the ushers rehearse the money-handling process to ensure security and accurate counting.

- In terms of processes, there are coordination, planning, scheduling of activities, and effective budget and financial management processes that need to be in place with good communication. In fact, processes should be so effective as to change the organizational culture of the church.

- In terms of equipment, these are more obvious. We have microphones, speakers, consoles, lights, and projectors. Cables that connect, stands that support, musical instruments that play. But we also use collection baskets, cups, and plates for communion, and in historic churches, other items like incense and censers, altars, and tablecloths, which are part of every Sunday worship.

If the church decides to change the worship service schedule to Friday night, it will indeed have an impact on all the other areas. But how would we know if the schedule change benefits the church ministry? We would need to assess the effects on various aspects.

Additionally, technology as an interconnected network exists within the broader context of culture.

For example, in Western employment culture, Friday night might not be an ideal time for a service. It could lead to decreased attendance, poor participation, and lower tithes and offerings. This, in turn, would have numerous other effects. Any technological decision in the church ministry should consider all five areas mentioned. Each process of technological maturity should seek a balance of these five areas in its growth. I will further explore this topic in **Chapter 7**.

On a separate note, if the church or Bible institute holds personal information of its members or students, it needs to comply with requirements like HIPAA, known as the *Family Educational Rights and Privacy Act* or FERPA.

Purpose of Technology

The German philosopher Martin Heidegger proposes the following two definitions of technology: the instrumental definition, which states that "technology is a means to an end," and the anthropological definition, which establishes that "technology is a human activity." According to Heidegger, these two definitions complement each other to reveal the essence of technology.

To that effect, Heidegger writes:

> The manufacture and utilization of equipment, tools, and machines, the manufactured and used things themselves, and the needs and ends that they serve, all belong to what technology is. [5]

In the previous section, I defined technology from a human perspective. In fact, the technology model I presented starts with people and continues with the human activities of systematizing, processing, and regulating the use of devices. These are the means to achieve an end.

However, the definition of technology would not be complete without considering the matter of the end that is pursued through those means. In other words, the purpose, the goal, or the "what" of technology.

The "what" of technology serves a tripartite or three-dimensional purpose. I like to refer to this tripartite purpose as the "trilogy of technology."[6]

The trilogy of technology consists of strategy, tactics, and intelligence.

The strategy of technology relates to the ultimate purpose, or the "why." The tactics of technology pertain to the operations in their context, or the "how" and "where." And the intelligence of technology involves measuring "when" the results are achieved and subsequent decision-making.

Intelligence also serves the purpose of refining the formula that allows for the replication of the combination of strategy and tactics to achieve the same results each time they are implemented.

I want to refer to each of these purpose dimensions in terms of the goal or the "what." And I will do so by following an example from the Bible.

The Gospels recount that on several occasions Jesus spoke to the crowds by the shores of a lake. For instance, Mark 4:1 reads:

> Again, Jesus began to teach by the lake. The crowd that gathered around him was so large that he got into a boat and sat in it out on the lake, while all the people were along the shore at the water's edge.

Other passages confirm this practice of Jesus (Matthew 13:1-2; Mark 2:13; Mark 3:7-9; Luke 5:1-3). When the people gathered to see Him, Jesus would get into a boat and move away from the shore. In this way, His voice would project over the surface of the water, and due to differences in air layers and the terrain's slope, He could be heard by the multitude. This technique has been studied as a method with clear acoustic purposes and has been compared to the design of a Greek amphitheater.[7]

The trilogy of technology consists of strategy, tactics, and intelligence.

So, the "what" of Jesus was that people would hear the message He wanted to communicate. This makes sense because hearing the message of Christ is what generates faith in peoples' hearts, as Romans 10:17 tells us.

Let us see how the trilogy of technology works in this context:

- In terms of strategy, Jesus had a much higher purpose in mind than simply being heard. Jesus wanted to communicate His message of the gospel of the Kingdom and share the good news of salvation. He wanted to teach the people and gain followers. In the passage from Mark 4, Jesus taught them the value of the word of God, which is sown in good hearts to grow and bear much fruit in the lives of those who practice it.

- In terms of tactics, we observe that Jesus' method was consistent. He would go to the lakeshore, choose a boat, have it pushed out from the beach, and begin instructing the people with parables and allegories. We know that He was able to gather many followers by listening to Him, including Matthew (Mark 2:14), Paul, James, and John (Luke 5:8, 10) whom He chose as His apostles.

- As for intelligence, we cannot guarantee that the evangelists used evaluation models to measure the results of the methods employed by Jesus. However, based on the passages I mentioned earlier, we can know that numerical growth was significant. There are references to "large crowds" (Matthew 13:2), "many who followed him" (Mark 2:15), and that he had "healed many" (Mark 3:10). He had the opportunity to relate "many similar parables" (Mark 4:33) and speak "many things" (Matthew 13:3) with the purpose that they would bear fruit and yield "a hundred, sixty, or thirty times" (Matthew 13:23).

In summary, the "what" of Jesus was that people would hear his message, the "why" was to teach them, gain followers, and heal them, the "how" was the methodology he used and the "where" was where he employed it, and the result occurred when the multitudes listened to him and followed him in great numbers.

In your ministry, all three components of technology should be present. The strategic element should precede the tactical one. This means that you must answer the "what" and the "why" before knowing the "how" and the "where." In other words, before deciding where to place a projector, you should define what you want to achieve with that projector and why you want to use it. Before deciding which system to use for managing the church membership, you should define the ministry purpose you intend to accomplish (such as providing better pastoral care or getting to know your congregation).

Once you go through the implementation process, you should evaluate to determine if what you did was worthwhile and produces the desired results (for example, you ask yourself if the text-based giving system has improved revenue or if the change in schedule has increased attendance and participation).

Every technology maturity initiative should address questions about its goals, articulate its motivations, define its procedures, and measure its effectiveness at each step. Your technology plan should consider strategy, tactics, and intelligence, all three.

Technology in Ministry

Understanding the purpose of technology is important not only as a complement to its definition but as the foundation for answering the question, "Why should I use technology in ministry?"

First and foremost, it is impossible to do ministry without technology. This has been true since ancient times. The Bible contains many examples of technology being used for ministry purposes, and I would like to mention some of them.

The most rudimentary forms of worship, which appear in the first book of the Bible, include the sacrifice of animals on an altar (Genesis

4:4). For this, humans had to use objects to sacrifice the victims, have developed mastery over fire, and construct stone altars.

The same chapter, Genesis 4, tells us that Jabal was the first to live in tents and raise livestock; Jubal was the creator of the first wind and stringed instruments; and Tubal-Cain was a blacksmith and "instructed all who work in bronze and iron" (Genesis 4:19-22).

In the wilderness, the people of Israel had to employ a series of technical skills to construct the portable tabernacle under the direction of Bezalel and Oholiab (Exodus 31:1-11; the description of these tasks occupies a significant portion of the book).

These examples from the Bible demonstrate that technology has always been intertwined with ministry. From sacrificial altars to musical instruments and the construction of the tabernacle, technology played a vital role in facilitating worship and carrying out God's instructions.

Therefore, when considering the use of technology in ministry, it is important to recognize its historical and biblical significance. Technology is not an optional add-on but an integral part of fulfilling the mission and purpose of ministry.

It is imposible to do ministry without technology.

God gives Moses the law written on the first tablets, which were created to read the word of God, and they are kept in a specially made case (Exodus 31:18; 32:15; 34:1-28). A mini replica evoking the Ark of the Covenant was to be carried tied on the hand and forehead and affixed to doorframes in what is known as a mezuzah, a practice that continues to this day. Adorned with the first Hebrew letter of

Deuteronomy 6:4-9 (*shin*), it contains the first great commandment as a reminder of the other nine.

Also in the Pentateuch, we read about how the institution of the Aaronic priesthood aligns with my organization of technology into five areas:

- **People:** Aaron and his sons were exclusively responsible for all priestly duties.
- **Systems:** The cycles of sacrifices, the burning of incense, the anointing, the unleavened bread, and the transportation of the sanctuary are microsystems of the tabernacle's activities.
- **Processes:** The washing of the priests and the offering of sacrifices during special feasts are detailed step by step, among other processes.
- **Policies:** The description of the regulations to be followed in worship activities is meticulous; the book of Leviticus delves further into them.
- **Devices:** There are many utensils, altars, chests, tables, vessels, censers, lamps, and garments. The Bible even provides a detailed inventory of these items.

King David reformed the worship system by making music an integral part of the worship to God. In addition to being a talented musician playing the harp (1 Samuel 16:16), he composed psalms and songs, formed a choir (1 Chronicles 23:5; 25:1, 6, 7), and invented various musical instruments (2 Chronicles 7:6).

The prophets used many technological references as metaphors to express the word of God. I have already mentioned how Jesus demonstrated mastery of acoustics and the pedagogy of parables. Likewise, the apostles were able to fulfill their mission thanks to the technological advancements in transportation and communication in the Roman Empire. It is well known that Paul, in addition, was skilled in the art of making tents, which enabled him to support his preaching throughout the known world.

As you can see, technology permeates multiple areas of biblical ministry. Due to space limitations, I cannot explore all the references to technology in the Bible; that is the subject of another book. However, I can use some of these references to outline some purposes of technology in ministry.

Based on the references I have cited technology is used for the following tasks in ministry; these respond to the strategic question of "Why?":

- Worshipping God and not worshiping technology.
- Attracting others to the message of Jesus.
- Gaining followers of Jesus.
- Teaching Jesus' followers.
- Serving the people of God.
- Healing the sick in body and soul.
- Organizing and financing ministry work.

We use technology in ministry because we desire that people "hear" the word of Christ through every possible means, so that they can worship God, be drawn to Jesus, become His followers, learn to live the Christian life, serve others, heal, and restore others, all while doing it in an organized and coordinated manner. This can take various forms in practice.

Technology permeates multiple areas of biblical ministry.

Rather than focusing on purchasing the most powerful audiovisual equipment, we should teach and foster worship in the true biblical sense.

Rather than building and maintaining a website, we should establish an impactful integrated digital presence to reach people.

Rather than sending sporadic invitations to special events, we should strengthen relationships with the local and virtual community.

Rather than communicating regular church activities, we should clearly express the vision of the leaders and stay connected with congregation members.

Rather than focusing on the successive production of weekly services, we should dedicate time to research online and through surveys to understand the needs of congregants and serve them more effectively, bringing healing and restoration.

Rather than centering all activities around the church building, we should decentralize the church through diverse small groups, including the possibility of exclusive online groups through video conferencing programs.

The important thing is to maintain the ministry purposes in their rightful priority, no matter how familiar they may seem. While these purposes may be simple, they can easily be lost sight of when we become engrossed in the daily tasks of implementing and managing technology.

We cannot lose sight of the ministry of the Word and its fundamental purposes. They need to be clearly articulated and kept in the right perspective.

Every technology maturity program must remain rooted in the fertile ground of ministry purposes to bear fruit. But first, we need an appropriate maturity model. Let us explore some maturity models in the next chapter.

MATURITY MODELS

Begin—to begin is half the work, let half still remain; again begin this and thou wilt have finished.

– DECIMUS MAGNUS AUSONIUS, ROMAN POET.

I was working as a consultant for a Christian University when I noted a great waste of resources. The high-performance central information storage system was being used to store personal documents and photos, much to the surprise of the department's directors. Institutional files of interest were the exception. Without delving into technical details, I want to put things into perspective. At the time of writing this book, an external hard drive with a capacity of 1TB costs less than $100. The same 1TB capacity in the high-performance system that the university had purchased cost around $10,000. That is a hundred times more! The complete system costs over half a million dollars.

It was as if they had bought a Lamborghini to drive in heavy traffic on the 405 freeway in Los Angeles. I approached the then infrastructure technology manager and told him that we needed to change the way we used the central storage. He agreed with me.

The high-performance system, with its fiber-optic hard drives, dedicated computers processing information at high speeds, and powerful network connectivity, should not be wasted on storing Paula's cat photos, which, by the way, she only looked at occasionally when Anthony showed her his dog's pictures.

This expensive system was designed as a platform for databases and operating system virtualization, which are always online, serving the university's administrative and educational data.

I started by writing a project proposal. To execute it, we needed to invest nearly $300,000 to make it a 100% redundant system. This meant that we needed double of everything. Two drives for every existing one, two power generators, two network switches, two network cards per server, two... two... two. We could not afford to have a single point of failure.

The project was approved, and we soon got to work. We acquired the necessary equipment and started transforming the central storage system into a 100% redundant system. Then we began migrating each computer that used local hard drives to the central storage. We started with small and less critical systems, migrating them one by one. Then we moved on to medium-sized systems and finally transferred the critical systems that handle all the university's data.

Student records, admissions, accounts payable, finance, payroll, and many others—our purpose was to virtualize and consolidate each system on the new platform. Our goal became to migrate each system to the new storage environment. To achieve the migration of each one, we implemented techniques that progressively required more commitment from us.

Goals are very important. Goals are milestones that allow you to recognize your progress. Goals focus and motivate you.

I kept a detailed record of all the systems that were being moved. Then I proceeded to implement installation, configuration, system monitoring, and issue notification standards.

With each system that was moved, the process accelerated. The migration of all the university's systems took seven years.

Goals are milestones that allow you to recognize your progress.

After more than a hundred systems, over forty databases, achieving 100% redundancy and virtualization, I can say that it was worth it. We were able to reduce the risk of failures to less than 0.1% and save the university hundreds of thousands of dollars in operational costs. We had undergone a process of technological maturity that began with a defined purpose and clear goals.

Maturity in Life

Although discussing models of technological maturity is quite common in the field of computer science, I am alarmed by the surprise of Christian leaders when this topic is mentioned. I often find myself trying to explain why it is necessary to mature and have a model of technological maturity to know how to mature.

Life has a model of maturity that we learn in basic biology: every living being is born, grows, reproduces, and dies. There are even more detailed classifications with all the stages of life. Everything that is alive, everything whose purpose is to bear fruit, needs a model of maturity.

For example, childhood consists of four stages: infancy, which lasts from 0-4 years old; early childhood, which includes children between the ages of 4 and 7; preadolescence, which spans from 7-12 years old; and adolescence or puberty, which covers the ages of 13-20.

Each stage has its distinctive characteristics and comes with its own age-specific challenges.

All these stages lead to the reproductive capacity of adulthood. Only adults can procreate. Only mature individuals can be parents in every sense of the word.

Maturity in Technology

The same applies to technology. For ministry technologies to unlock its power for fruitful ministry, you need a maturity model. Having a model of technological maturity is essential for growth. A maturity model puts us in perspective, allows us to know where we are, and motivates us to reach our full potential.

A business adage says, "What gets measured is the only thing that gets managed." A crucial component in developing a maturity model is being able to measure progress. A good maturity model will let us know where to start and where to continue. It will provide us with the roadmap markers - the goals - that will allow us to recognize and even celebrate our progress.

There are several maturity models that stem from the study of technology. In this section, I want to review four models while keeping these four purposes in mind.:

1. To inform you about existing models in the industry. With this, I aim to expand your knowledge of the subject and expose you to a new dimension of possibilities.

2. To provide you with a common vocabulary. When I used to train at the technical school, I would tell my students, "50% of learning a new technology is learning its vocabulary." I do not have a scientific way to prove that it is exactly 50%, but I believe you get the idea.

3. To observe how each model defines the process of maturity. Maturity as a process is common to all these models because each step brings you closer to the reality you want to see; a way to "predict the future' if you will.

4. To distill the five principles of technological maturity that are common to each of these models. The principles are universal and apply to the technological maturity of ministry.

The first model comes to us from William L. Nolte's book, *Did I Ever Tell You About the Whale? Or Measuring Technology Maturity*. This book is, if not the most comprehensive, one of the most comprehensive ones I know on the subject.

Nolte proposes an intuitive biological metaphor of seven stages to achieve technological maturity: Conception, birth, childhood, adolescence, adulthood, maturity, and old age. Considering that some technologies tend to decline when they become obsolete, he then adds two more stages: senility and death. [8]

The advantage of this model is that it uses a familiar process. We all understand this cycle of life and are familiar with these transitions. Nolte applies it to technology. Every technology begins with the conception of ideas, its birth when it is brought into production for the first time, its growth when it is adopted, and reaches its maximum capacity in maturity.

Another model, and one of the most utilized in the software industry, is known as the *Capability Maturity Model* or CMM.

CMM was created as a response by the Software Engineering Institute at the prestigious Carnegie Mellon University. It establishes the concept of software process maturity as:

> ...the extent to which a specific process is explicitly defined, managed, measured, controlled, and effective. Maturity implies a potential for growth in

> capability and indicates both the richness of an organization's software process and the consistency with which it is applied in projects throughout the organization.[9]

Note some important things. Technology maturity has two dimensions in this model. The first is growth in quantity: growing in the number of capabilities. The second is growth in quality: consistency and reproducibility, better programs, or a better ministry experience. I discussed this in **Chapter 2**.

Capability maturity involves progress in the capabilities of software. Maturing software can deliver optimal results. According to this model, there are five levels of maturity and four controlled processes that help you transition from one level to another:

1. **Initial**
   With disciplined processes, it transitions to...
2. **Repeatable**
   With consistent and standardized processes, it transitions to...
3. **Defined**
   With predictable actions processes, it transitions to...
4. **Managed**
   With continuous improvement processes, it transitions to...
5. **Optimized**

The benefit of this model lies in its emphasis on management, evaluation, and control. Your ministry technology programs will not mature on their own accord; they require your explicit involvement to define maturity, how you plan to achieve it, and what results you aim to obtain.

The *Diffusion of Innovations* model proposed by Everett Rogers, represents innovation as a maturing evolution in six steps[10]:

1. Problem or necessity

2. Basic Investigation
3. Solutions Development
4. Product Commercialization
5. Diffusion and Adoption
6. Consequences or Effects

Note that in this model, there is a recognition of and interest in the degree of transformation that the maturity of an innovation could cause in the social environment. It focuses on the effects of innovations on human behavior.

The stages in these models do not necessarily follow an exact order. Sometimes they can be interchanged or overlapped. The important thing is that you have, understand, and follow the model so that you can observe progress, allow the controls to provide feedback for adjustments, and achieve the desired end more frequently.

The main value of this model is innovation. It helps us connect a problem or a need with the sequence of stages that lead to technological innovation.

You have certainly heard the saying, "Necessity is the mother of invention." In fact, the phrase comes from Hugh of Saint Victor, a Saxon priest, theologian, and pedagogue from the Middle Ages, who wrote: "Necessity is the mother of technology," [11] circa 1100 A.D. There must be a direct connection between your ministry technology and the needs you want to address; otherwise, there would be no maturity. Technology should meet a ministry need, and it will do so better as it matures.

The last model I would like to mention is the *Kerzner Approach®* for project management. He writes:

> Maturity in project management is the implementation of a standard methodology and accompanying processes such that there exists a high likelihood of repeated successes. [12]

This model establishes five maturity levels:

1. Common Language.
2. Common Processes.
3. Singular Methodology.
4. Benchmarking.
5. Continuous Improvement.[13]

Kerzner applies the five stages of CMM to Project Management in his book of the same name. He acknowledges that project management is a purely human activity, albeit assisted by technology, and that the ultimate result may be technological products.

A maturity model does not regulate the functioning of technology itself, but rather human behavior.

Once again, a maturity model does not regulate the functioning of technology itself, but rather human behavior. It is we who need to mature in our interactions with technology. That is why our vocabulary, our ways of doing things, our methods, our ability to assess maturity, and our observation of real-world results all need to mature.

The benefit of this model lies in its emphasis on people as recipients and managers of maturity, as I mentioned in the previous chapter.

Therefore, an appropriate maturity model is, according to Nolte, intuitive, progressive, and helps you understand growth; according to CMM, controlled and evaluated; according to Rogers, innovative yet rooted in need; and according to Kerzner, for and by the people.

All these elements must be present when developing a maturity model for ministerial technology to make it appropriate and relevant. The individuals responsible for technology in the ministry, including the pastor and their leaders, need to follow an intuitive, progressive, controlled, evaluated, and innovative model.

Parallels

I understand that I have been discussing the topic of technological maturity, but I am also very interested in the parallels that exist with individual and institutional maturity. The reason is that maturity also applies to the Christian life and the development of the church ministry.

I believe that when we find parallels, they affirm both the need and validity of a maturity model. That is why I want you to consider for a moment the maturity of your calling: Your own calling also goes through stages of maturity.

Jeff Goins skillfully develops this topic in his book *The Art of Work*. He acknowledges that the calling or vocation, finding that unique thing for which you were put on this earth, or what he calls "your life's work," goes through a process of various stages.

He writes on this regard: "There is a process to finding your life's work. Although at first it may feel chaotic, there is an order emerging from such chaos." [14]

So, your calling begins with chaos and culminates in a valuable work of art. To develop your full potential, the calling goes through the following seven stages: [15]

1. Awareness
2. Apprenticeship
3. Practice
4. Discovery
5. Profession

6. Mastery, and
7. Legacy

Finding your calling is not a one-time event that, if you miss it, it is gone forever. Finding your calling is a gradual process. Discovering and fulfilling your calling is a journey of maturity. Furthermore, note that the highest expression of your calling is the ability to leave a legacy for others. This once again highlights the capacity to bear fruit and reproduce, as explained in **Chapter 1**.

Figure 1 below provides you with a comparative table showcasing the similarities and differences between these models.

Maturity Model	Biological Model	Capability Maturity	Diffusion of Innovations	Project Management	Calling Maturity
Proponent	Nolte	Paulk	Rogers	Kerzner	Goins
Stages	Conception	Initial	Necessity	Language	Awareness
	Birth	Repeatable	Investigation	Processes	Apprenticeship
	Childhood				Practice
	Adolescence	Defined	Development	Methodology	Discovery
	Adulthood	Managed	Marketing	Benchmarking	Profession
	Maturity	Optimized	Diffusion	Improvement	Mastery
	Old age		Effects		Legacy

Figure 1: Comparing Maturity Models

At first glance, you can notice a few things:

1. The most comprehensive models have seven stages.
2. The initial stages are for identifying the need, conceiving ideas, and establishing a common language.
3. Maturity is associated with the optimal state, mastery, and dissemination or reproducibility, as mentioned before.
4. Maturity requires planned management or intervention; it is not left to chance.
5. Goals are clearly defined in a maturity process, as the next level of maturity becomes the goal of the previous level.

Maturity Principles

There are many other maturity models available. A Google search for the phrase "Technology Maturity Model" returns over eighty million results. Each model adapts the principles of maturity to a specific industry or common human enterprise. To date, no model has been specifically tailored to the maturity of ministerial technologies.

One of the brilliant aspects of Nolte's work is that he takes some of these technology models and integrates them into his biological model, providing a more comprehensive perspective while also identifying some deficiencies in existing models.

Nolte points out that most models focus on the initial stages of technology, but few have addressed maturity in the more advanced stages. [16] The roadmap presented in the next chapter puts into perspective and balances all stages of technological maturity in ministry.

Furthermore, we have studied the biblical foundations of maturity (**Chapter 1**) and some technology maturity models (**Chapter 4**). By analyzing and aligning the Bible with these models, I realize that there are principles that govern every maturity process, including technological maturity. Now we can review these principles that govern maturity by incorporating both sources.

The principles of technological maturity are:

The Substitution Principle

Maturity is achieved by replacing growth alternatives where there are negative attitudes. It is about living with aspirations. The best way to mature is by substituting things you do in the present for things typical of the next stage.

Generally, we do a better job of embracing the new than letting go of the old. Substitution allows you, for example, to use your website and email system as the primary mode of communication to replace the 10-15 minutes of announcements during the worship service. Another idea could be to pre-record the announcements in a video, prompting ministries to think ahead about what they want to communicate.

The Transformation Principle

Maturity seeks to change human behavior in its interaction with the environment. Technology is one of the primary ways that humans interact with the material world, which is why technology is so transformative.

When we adopt a new model and discontinue an old model, there are repercussions in other areas of technology. People must change their habits, systems need to be redesigned, processes and policies need adjustment. Devices are decommissioned, and new ones emerge on the scene. Continuing with the previous example, perhaps an email marketing system is needed instead of installing speakers in the hallways.

The Revitalization Principle

Maturity revitalizes. Giving life is a unique characteristic of living beings. Technology itself does not have the ability to produce life, but it can promote, protect, and prolong it. That is why technology can be a vital agent.

It is evident that technology functions as a vital agent in cases of healthcare technology: a pacemaker, a blood pressure monitor, or a respirator sustain life. In your ministry, there will also be essential technology. It is your task to identify how technology in your ministry fulfills a vital role.

The Reproducibility Principle

Maturity aims for reproduction, bearing fruit in character, quantity, and quality. Technologies should create mechanisms that foster creativity, multiplication, and excellence in the functions of ministry.

While this may be obvious in our minds, it is not always evident in our practice. Sometimes we become attached to inefficient technologies that limit our reproductive capacity. Certain technologies facilitate reproducibility by promoting collaboration, communication, and teamwork through productivity tools.

The Spirituality Principle

Maturity is guided by the Holy Spirit and has spiritual purposes. As we have seen, spirituality has always been mediated by technology from ancient biblical times to the present day. Technological maturity can facilitate and foster spiritual maturity to assist you in your devotion, finding your purpose in life, your calling, and your service to others.

There are various mobile applications, for example, that can guide your spiritual growth, help you adhere to certain spiritual disciplines such as prayer, Bible reading and reflection, instruct you in the doctrines of the Christian faith, and even connect you with accountability groups in areas of temptation or weakness.

Being spiritual is being mature.

These principles are not principles of every technology. Technology itself cannot be spiritual, for example. The principles apply to the entire maturity model, so technological maturity can incorporate them.

The five principles form a hierarchy that ranges from the foundation of substitution to the pinnacle of spirituality. Substitution produces transformation. Transformation produces revitalization. Revitalization produces reproducibility. And reproducibility produces spirituality.

Think about this for a moment: Is it possible to become a mature person without substitution? This is initially what repentance is about, substituting the focus of your life to shift from being self-centered to Christ-centered.

You can see the substitution principle at work in Galatians 2:20, which reads:

> I have been crucified with Christ and I no longer live, but Christ lives in me. The life I now live in the body, I live by faith in the Son of God, who loved me and gave himself for me.

Well, the same applies to technological maturity. You can focus your efforts on the technology itself, or you can center them around the mission of Jesus Christ. After all, focus is everything because if you see where you are going, you will go where you see. If you shift your focus toward spiritual life, you will pursue it.

Any model of maturity for ministerial technologies must therefore include these five principles, which have a strong biblical foundation while mobilizing technological initiatives in a virtuous circle, a cycle of continuous improvement to bear much fruit.

To date, a model of maturity for ministerial technologies has not been developed, that is, a model that applies maturity principles to the way the church and Christian ministries use technology to promote their mission. Today, I want to address this gap. This is the topic of the next chapter.

MATURITY ROADMAP

Each of us creates a representation of the world in which we live – that is, we create a map or model which we use to generate our behavior. [17]

– BANDLER & GRINDER, FOUNDERS OF PNL.

New innovations, whether incremental or transformational, are developed: they don't just happen. [18]

– CURTIS CARLTON & WILLIAM WILMOT.

I have fond memories of my youth camps. The friends I made and the things I learned. My encounters with God. All the experiences were enriching. But nothing triggers my memories more than the games we played.

I remember how much fun we had playing "The Treasure Map." The youth would organize into teams called patrols and we would be taken to a secluded area of the game zone. The facilitators would prepare the terrain with a series of obstacles of varying difficulty. They would place objects that we needed to find, some on tables, others stuck to trees, and others buried in the ground. They would also rope off the more challenging or steep sections of the terrain to guide us.

The purpose of the activity was to complete the course in the shortest amount of time possible and reach the end to find "the treasure" without losing any team members. Sounds easy, right? But there was a catch. All members of the patrol were blindfolded, except for one. And that one who could see was restricted from viewing the terrain. They could only see... the treasure map.

The one who could see the map was the captain of the patrol. Their task was to ensure that the whole team stayed together while giving instructions on how to proceed based on the map.

It was hilarious to see how the teams blindly followed the captain and the different techniques they came up with to stay united. After the game ended, we would have time for reflection and learn the principles of the activity.

The Leader Follows the Map

A leader is someone who leads their organization to achieve its goals. A leader is someone who looks out for the interests of their team. A leader is someone who has a vision for the ministry. A leader is someone who sees the map.

Many of those we call leaders in our churches and Christian organizations do not see the map. The ones who do may see the map of their vision and even of the mission of the church. But they do not see the roadmap of technology maturity and, therefore, miss out on many opportunities that technology offers the ministry. I do not blame them. Until now, a technology maturity roadmap for the ministry has not been developed.

The absence of a technology maturity roadmap for ministry is a lamentable condition, considering what the Bible says about maturity. Could this be the reason why many ministry technology programs are infants and incapable of bearing fruit? Could this be why many of our leaders do not understand the advantages of ministry technologies?

It was not enough for me to realize this. I was eager to be part of the solution. The model I am going to present below is the result of twenty years of experience in the field of information technology.

God has allowed me to refine this model by serving as an agent of technological transformation in dozens of universities and Christian education institutions.

I have had the opportunity to apply this model to Christian ministry in a variety of settings with excellent results. I was able to guide them to see the map and find the treasure.

This chapter explains the roadmap of ministry technologies. It consists of seven steps to unlock the power of technology for fruitful ministry. But before explaining each of these seven steps, I want to review the necessary requirements to achieve technological maturity. These requirements should permeate each of the steps as you progress towards a greater capacity to bear fruit.

Maturity Requirements

Experience is the best teacher in this regard. Working as a technologist has shown me that moving an organization from one maturity stage to another is hard work. Technological maturity is much more complex than it appears at first glance and has requirements that I would like to mention. The following requirements are indispensable to achieve technological maturity.

Clear Vision of the Future

The technology maturity roadmap aims to stimulate your thinking about technology so that you can create a clear vision and understand the role that technology plays in achieving that vision.

As a leader, it is imperative that you define, write, and communicate the vision, not for the next two months or the next year, but for the next three to five years. Do not talk about it until you are clear about it and have it in writing.

Strategic Alignment

How does the popular saying go? "There's a long way from saying to doing." Well, from vision to mission, it's the same. Strategic alignment means that what the church sees (vision) aligns with what the church does (mission). Strategic alignment means coherence between message and purpose.

It is common to find institutions with disjointed strategies. They are disjointed from the vision when they are headed in a different direction from where they want to go or say they want to go. They are disjointed from the mission when what they do does not align with what the Bible says they should do or with the vision they want to achieve. They may even be disconnected from their organizational identity, claiming to be one thing, but their actions do not reflect it.

As a leader, it is crucial for you to generate congruence between your identity, vision, and corporate mission. This congruence should be transferred to everything your organization does. Your values, goals, logos, and other visual aspects, your website, programs, and projects—everything should be conducted in complete alignment.

Good Project Management

When strategic alignment is established, several goals will emerge, which in turn will generate various projects. Projects are concrete initiatives with a defined time frame that will produce a deliverable, a product, or a result. There are projects in ministerial technologies and there are ministry projects with a strong technological component. Some projects may even last for several years. In any case, each project must be carefully planned and controlled with great attention to detail.

As a leader, it is crucial for you to establish selection and initiation criteria, execution and control procedures, and evaluation and closure measures for your projects. There is an entire discipline for project management that I recommend you become familiar with.

As a leader, it is crucial for you to generate congruence between your identity, vision, and corporate mission. This congruence should be transferred to everything your organization does.

Good Change Management

I once heard someone say that God loves us as we are, but He loves us too much to leave us as we are. The Gospel is a transformative journey. Every journey towards maturity is a journey of change. However, I am surprised by how little our leaders understand about change management, and there is so much to learn in this area.

As a leader, you need to be an active participant and an agent of change. Successful implementation of directed changes is an essential component of your leadership. As you will see in the rest of this chapter, effective change management will help you transition from one stage to another in your process of technological maturity.

Educational Intention

I am indebted to my good friend Jairo Ospina, who also assisted me as a beta reader for the Spanish version of this book, for his understanding of this requirement. For many years he has worked as a technology project manager for several educational institutions. When reviewing this topic, he added: "every major technological initiative demands changes in organizational culture, and every change requires a certain level of educational effort." In other words, every process of cultural change requires re-educating people.

As a leader, you are an educator. You must be intentional about it. Education is your ally in changing organizational habits, transforming the working culture, fostering productivity, and encouraging those you lead to mature.

Continuous Evaluation

Evaluation is what keeps you accountable throughout the entire maturity process. Evaluation, in addition, ties together all the other requirements. I mean that evaluation will allow you to know if your vision is clear and has been understood by the members of your team, organization, or congregation. Evaluation will show you if your strategy is aligned, if your projects are completed on time and within budget, if the change achieved is desired and a result of your management, and if your educational initiatives enable people to grow.

As a leader, it is crucial to regularly evaluate each step and analyze the results of your evaluation to make informed decisions. I have dedicated **Chapter 8** to the evaluation of maturity.

7 Steps for Fruitful Ministry

The following maturity roadmap consists of seven steps. Each of these seven steps towards technological maturity is like a growth mark on the door lintel or a milestone on the road. They are stages of development or stages of progress.

Your technological initiatives can begin to mature through planning. They can show evidence of growth by taking these steps or going through these stages. You can even measure growth tangibly if you take the time for evaluation.

At each step of this model, I will incorporate progress indicators. These will help you visualize your growth so that you can, when the time comes, celebrate your achievements.

Step #1: Initiate

Initiate is to begin. It is the starting point or the first stage. It is when you make the decision to start. The beginnings of any technology implementation are characterized by four elements:

- Ignorance,
- Indecision,
- Incompetence
- Innocence.

We should not let any of these elements make us feel bad or underestimated. I know that ignorance or incompetence generally have negative connotations, but it should not be the case, and I assure you that I do not assign that negative connotation in this context.

Ignorance says: I do not know what to learn. Indecision says: I do not know what to expect. Both are normal components of any beginning. It is a stage of uncertainty.

For example, perhaps before starting to read this book, you were unaware of technological maturity. You were in ignorance. Now you know that there is such a thing as technological maturity. Although you are not an expert in technological maturity processes, you are no longer in ignorance because you now know that such a thing exists. Now you can be proactive and, if it interests you, mature in that knowledge.

Similarly, you may take the first steps in the process of technological maturity and not know if it is worth the effort or how it will turn out because there are many uncertainties, but you will gain confidence as you progress. Thus, the unknown becomes known and the uncertain becomes certain.

Old Hispanic folks often say, "Nobody is born knowing." We must overcome ignorance and uncertainty if we want to bear fruit and endure. We overcome ignorance and indecision in this initial stage.

Escaping ignorance is essential to avoid an unhappy ending. That is why God Himself said:

Hosea 4:6: "my people are destroyed from lack of knowledge."

If you already know that you do not know something, but you are aware that you do not know it, it is no longer ignorance. You can now act. If you take the first steps and realize that the model supports you, you affirm your conviction that it will also support you in the following steps.

In the initial stages, decisions not to learn and not to practice must also be made. The decision not to learn about a subject is not ignorance but incompetence. The decision not to practice something is also not ignorance but innocence.

Incompetence says, "I know that I do not know, but I am not interested in learning." Innocence says, "I know that I do not know, but I am not interested in practicing it." Both are more mature components of this initial stage because they involve decision-making.

It is okay if you are not interested in knowing about a certain topic if that topic is not aligned with the maturity you want to achieve. The incompetence that results from a decision not to learn is mature.

You also do not want to be a teacher of what is wrong. The apostle said:

Romans 16:19: "I want you to be wise about what is good, and innocent about what is evil."

In other words, it is okay to remain innocent in doing what is wrong. It is okay to remain incompetent in other areas that do not lead us to the maturity we seek. It is okay to remain incompetent at higher levels of maturity, as long as it is an intentional decision and allows us to be effective ministers.

Ignorance says:
I do not know what to learn.

Indecision says:
I do not know what to expect.

Incompetence says:
I am not interested in knowing.

Innocence says:
I am not interested in doing it.

If your technology program is in this stage, you need to answer four questions:

1. What do we need to learn?
2. What causes us indecision?
3. What do we not need to learn?
4. What are we not going to do?

If you can provide comprehensive answers to these questions, your beginnings will be more solid, and... you will have traveled half the way.

The initial stage brings about feelings of indifference and even stagnation. We may not initially recognize that we are in this beginning stage, but we start to notice that our technology is aging—that our technical skills are falling behind, systems hinder rather than facilitate, processes become slow, policies are inadequate, and devices are outdated.

Finally, we come to realize that our ministries are not advancing, and we need a change. We may have been indifferent to others' feedback or the needs of those we are called to serve. While they are immersed in a digital culture, we remain in an analog culture.[19] Simply becoming aware of this moves us towards the initiation stage.

We can mature beyond the initiation stage by exposing ourselves to the unknown. Learn. Keep reading, surround yourself with the right people, and attend conferences and workshops. By doing so, you eliminate the possibility of becoming stagnant and transition from ignorance to the next step.

Step #2: Inquire

Inquiry inspires learning. It is the state that follows the initiation. It is when we are moved by curiosity and a hunger to learn, even though we may still be distant from new technologies. During this time, we may not be able to fully leverage all the opportunities that technology can offer because we do not know how. It is a time for research and reading. We seek resources and opinions. We ask others what it is all about.

The process of inquiring is accompanied by feelings of curiosity and shyness. We may not recognize that we are in the stage of investigation, but we do notice that certain technologies attract us, and we ponder the possibilities they can bring to the ministry. We may feel timid to step in or fear the unknown or the outcomes.

If you do not fail,
you have nothing to overcome.

You can mature from the stage of inquiry by making the decision to keep learning and abandoning the *status quo*, as I mentioned in **Chapter 2**. Do not let the mindset of "we've always done it this way" hinder your desire to explore and educate yourself. Take the knowledge you have acquired, whether it is a lot or a little, and put it into practice as soon as possible. Do this consistently and be committed, and you will move from this stage to the next.

Step #3: Investigate

To investigate is to dare to experiment. It is the state that follows the departure from intrigue. We give ourselves permission to try, and in doing so, we realize that we can indeed do it.

Along the way, we discover that the field is much broader than we initially thought and make decisions to continue.

During this stage, chaos reigns; it is a time of trial and error. It is okay to make mistakes and adjust. Do not let the fear of failure paralyze you. Give yourself permission to fail. We live in a culture that is afraid of failure, but the only way to make progress is by overcoming failures. If you do not fail, you have nothing to overcome.

The investigation stage brings about feelings of dissatisfaction and sometimes even rejection. You may not recognize that you are in the investigation stage, but you do notice that you are not satisfied with what you have learned so far or the results you are achieving. It is even possible that you are considering rejecting technology altogether and abandoning your initiatives. That's normal. You are not alone in feeling this way. This is when you become intentional about your maturity.

Intention is intentionally matured through a plan.

Intention alone is not enough. It requires intervention to move to a new level of maturity. Intention is intentionally matured through a plan. This is the stage where some planning is incorporated. If your church or ministry has plans to use technology in ministry, you are in this stage. Plans, not wishes. Wishes would belong to the previous stage.

It would be very helpful in this stage to start establishing effective project management methods as I described in the requirements, although typically these are developed later in the maturity process.

You already know that intention alone is not enough. Action is necessary. The first action to take is planning. The second is to follow the plan. Remember? Begin and continue to move from intention to the next step.

Step #4: Implement

Implementing is executing your plan. In this stage, we understand the possibilities of technology and create concrete solutions to real-world problems. We move from proposal to response.

It is the time when we form strategic partnerships with other resources to accelerate utility and results. Since you have already overcome the initial ignorance (and have decided in which areas you prefer to remain incapable) and have enough experience to overcome your failures, you can confidently identify those resources you need to complement your plan.

There is nothing wrong with hiring temporary staff to implement your plans. It is okay to consider hiring a ministry technology advisor or technology project management consultant. It is okay to hire developers and consultants.

Implementation is an appropriate opportunity to create standards and systematize your processes. Take the time to write policies and procedure manuals. We cannot underestimate this work as it is the foundation for the subsequent stages of technological maturity in ministry. As you will see in **Chapter 7**, this is one of the most deficient areas of technology in the church.

Implementation produces feelings of fulfillment and fascination. We may not recognize that we are in the implementation stage, but we do notice that we feel satisfied with certain achievements and are irresistibly drawn to tangible results. Achievements fuel new goals.

It can also generate false expectations. This can happen when we believe that we can solve everything with technology. Technology does not solve everything. Technology has its limitations.

Achievements fuel new goals.

In many instances, those limitations give rise to other technologies. If you find yourself at the frontier of the capabilities of a certain technology, document them in detail. It is possible that your creative power will be perfected in the weakness of the current technology. You can act on it if you mature to reach Step 7.

Many successful ministries remain at this level, but you can mature from implementation to the next level through reflection. Someone rightly said that good is the enemy of best. You cannot settle for what is good when there are others waiting for you to deliver what is best. Let your early successes propel you to reach the ultimate challenges. We do this by taking the implementation experiences of technology into the realm of ideas, lessons learned, and productivity improvement processes.

Step #5: Ideate

Ideation is about thinking. In this stage, you develop patterns and models based on your own past experiences.

During this stage, the ministry becomes more flexible and determined. Clear priorities are established. It is recommended that, if it has not been done yet, you initiate a strategic technology planning process that leads to defining concise goals and how to adapt and project into the future.

Ideation is a stage of profound and abundant creativity. You start creating your own identity to distinguish yourself from others. You generate many new ideas and unique, specialized applications of existing technologies. You propose new frameworks of reference. Now you become the expert in the field and serve as a reference point for others. Your ministry advises and helps others navigate their own maturity processes; you function as a mentor to the new generations.

Ideation produces feelings of adaptability and hope. We may not recognize that we are in the ideation stage, but we notice that we easily adjust to the changing conditions of the ministry and respond swiftly to updates and the surrounding culture.

Experience and flexibility are the fuel of hope; we know that God has helped us thus far and will continue to help us through the next change.

Ideation can be matured by incorporating elements of ministerial technologies into the identity of the ministry. It is one thing for a ministry to use technology, and another thing for technology to define the ministry, as we will see in the next step.

Step #6: Identify

Identifying is about distinguishing oneself from others and from other ways of using ministerial technologies. Identifying is the beginning of the path towards innovation.

Identifying can be understood in at least two different ways. First, in the sense of pointing out or indicating, such as when you identify a need or point out a direction. In this sense, identifying corresponds to each of the previous stages, as you need to identify what you are going to initiate, how you are going to inquire, what methods to attempt, which technologies to implement, and which ideas to promote.

The second sense is the creation of one's own identity. It is when technology becomes an essential component of your unique identity. The way in which you use technology distinguishes you from other ministries.

This stage refers to the creation of a corporate and ministerial identity that intentionally embraces technology to fulfill the mission. For example, when you go through a process of digital identity and presence or infuse administrative functions of the church with high-productivity technology to the point where you cannot conceive the ministry without these elements. The ministry gives its entire purpose to technology, and technology gives identity to the ministry.

Every ministry should define a digital identity and presence strategy. Digital identity and presence refer to everything that can be found on the Internet about your ministry. This includes your website, search engine results, email communications, and all your audio, video, and social media channels. **Chapter 9** applies the maturity concepts to the digital presence of a fictitious ministry.

Digital presence requires the design and collaborative effort of your team to develop a distinct identity and consistently communicate it in everything the ministry does, both online and offline.

Every ministry should define a management strategy that emphasizes efficiency and productivity. The term "back office" is used to encompass all the invisible activities necessary to run the ministry itself. Collecting tithes and offerings is the visible part, but financial management is part of the back office. Taking data from people who respond to the altar call is the visible part, but storing personal data in a software application is part of the back office.

High-productivity technology refers to back-office software applications that facilitate the creation and sharing of information, collaboration, effective communication, time management, data storage, and project management.

Effective management requires planning and good change management to create a distinct corporate culture.

Apart from having a well-defined and established identity, certain technologies become embedded in that organizational identity and become part of the ministry's DNA. At this level of maturity, technology becomes such an integral part of the ministry that certain ministries cannot be carried out without their respective technologies, as I mentioned in **Chapter 3**.

Identification produces feelings of unity and identity. The organization becomes more complete, and the data becomes more integrated. Technology becomes more stable and indispensable.

Standards are created, and processes are automated to stimulate the faithful reproduction of results.

We may not recognize that we are in an identification stage, but we do notice that the ministry is adversely affected when technology does not function as expected or when there are technical failures. When this happens, the "flow" gets disrupted, or unity gets disturbed.

An example of a mature ministerial technology at this level is the use of the printed Bible. The printed Bible allowed for the creation of standards and the ability to reproduce it accurately. The book of the Bible integrates all its parts and does not fail to function when you need to read it. When you see someone carrying the Bible in their hand or under their arm, you immediately identify them as a Christian or "people of the Book." Due to this deep identification, the Bible has become synonymous with Christianity.

Effective management requires planning and good change management to create a distinct corporate culture.

One can mature from identification by creating something new. Most ministries today do not make that leap. The electronic Bible, to continue with the same example, has made that transition to step #7.

There are few pioneering ministries when it comes to novelty. But it has not always been that way. There have been times in history where the Church has been a stronghold of innovation for ministry and for the broader culture. That is why the highest level of maturity is innovation.

Step #7: Innovate

Innovation is about originating, inventing, and even improvising. Innovation is about creating transformative technology. All efforts revolve around the mission through technological innovation.

Ministerial technologies are not only part of your demand but also part of your offering; the ministry is producing unique technology for other ministries and even for consumer society. We become centers of innovation.

When a ministry moves from problem to solution, it progresses from step #1 (Initiate) to step #4 (Implement). When it expands the possibilities of this solution and changes the ministry culture in such a way that the implemented technology becomes an integral part of what it does, it has moved to step #5 (Ideate) and #6 (Identify) respectively. But when it turns the solution into a concrete product capable of helping others, it is innovative. It has reached step #7 of the maturity process.

Innovation brings about feelings of novelty, purpose –when we realize that what we have is not just for our own benefit, and compassion– because of our desire to help others overcome similar problems.

We may not recognize that we are in the stage of innovation, but we do notice that we are producing programs or models that did not exist before, which others begin to imitate.

Innovation is about creating transformative technology.

Innovation has two sides. On one hand, it involves creating something new, and on the other hand, it involves eliminating the old. In both cases, any innovative organization needs to reinvent itself, whether to embrace the new or to adapt to the loss of the old. Both require creative processes.

In summary, this book is a call to innovation: Let us transform our society with the gospel by presenting it in new ways, as some old ways are no longer effective. Let us become innovators. **Figure 2** on the next page represents the complete maturity map.

I want to add something more about innovation: Innovation is cyclical.

Due to this characteristic of its nature, when you innovate, you embark on another journey into the unknown. In a way, this puts you back in step #1, as there are many things that have not yet come to light or that you will be exploring. You will start gathering the information you need and discard what you consider to be unimportant.

Every new endeavor puts you at the beginning of the maturity map, but you will not be in the same place where you started; your circumstances will have changed, the people you will be serving will be different, the context will change, your experiences will have matured, and your transformative impact will be greater.

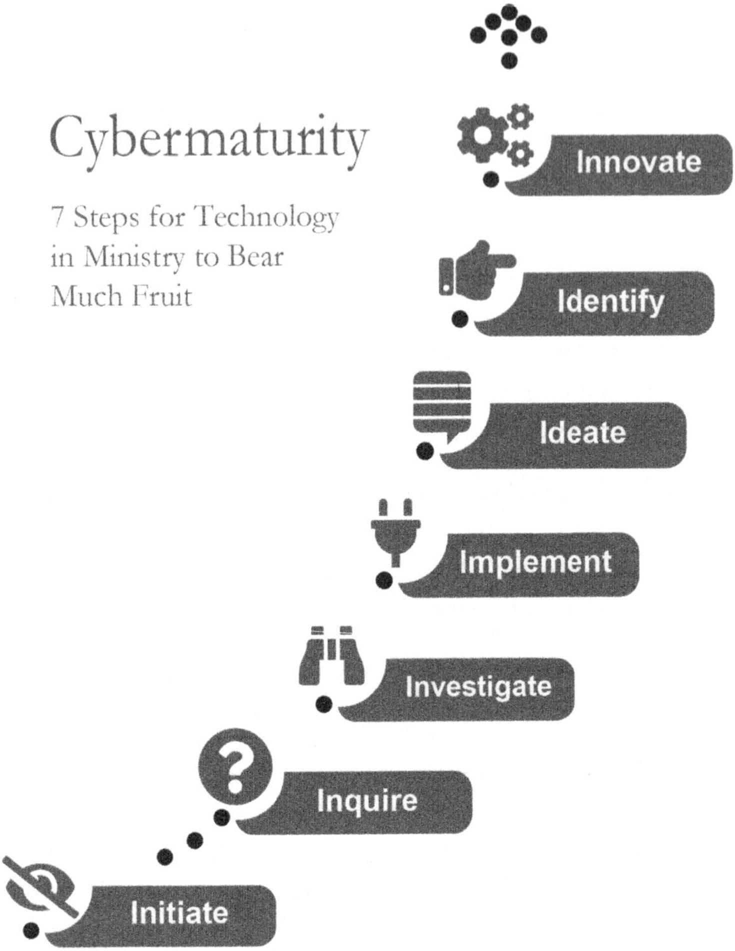

Figure 2: Maturity Roadmap

The cycle of innovation is not a closed circle but rather an upward spiral where with each complete turn, you ascend to a new level.

King Solomon realized this cyclical nature of innovation, but his downward spiral of sin and disobedience left him with a cynical and pessimistic attitude. We see this when he wrote:

Ecclesiastes 1:9: "What has been, will be again, what has been done will be done again; there is nothing new under the sun."

History has proven time and again that "the wisest man who ever lived" was mistaken. Many new and productive things have happened since the time of Solomon, such as the glorious birth of his great-grandson, the Messiah.

Solomon did not understand the innovative character of God, but the prophet Isaiah, as we will see in the next chapter, did. Innovation represents the very character of God.

GOD'S MATURITY

Men are like wine; some turn to vinegar, but the best improve with age.

– MARCUS TULLIUS CICERO, ROMAN PHILOSOPHER.

It was a warm spring night in 2007. After the garden and the themed gift shop, we walked through the hallway that led to the cellar. In a matter of minutes, we were seated in the dim candlelight, surrounded by oak barrels and fermentation columns.

This was not your typical business dinner. The hosts welcomed us with excitement and shared the story of their family, spanning generations of winemakers. Then, they introduced the chef, who explained the evening's menu and the careful pairing of each wine with every type of food.

Amidst compliments and acknowledgments, they presented the connoisseur. The expert chemical engineer delved into the art and technology of turning grape juice into the elixir of the gods.

My mind wandered. The chemical engineer in me wanted to hear every detail from the expert. But the theologian in me could not ignore all the biblical references to wine flooding my thoughts.

I recalled the first passage that mentioned wine when Noah planted a vineyard, got drunk, and ended up naked before his son Ham, the father of Canaan (Genesis 9:20-27). I also remembered that Jesus' first redemptive miracle, turning water into wine at a wedding, happened in Cana, the region named after Noah's grandson (John 2:1-11).

Canaan was cursed by Noah's words, but it was blessed by Jesus' miracles. In any case, the master of the banquet was pleased. It was customary for good wine to be served at the beginning of the wedding, and when the guests were already intoxicated and could not tell the difference, cheaper wine was served. But at this feast, the opposite happened. Jesus was not someone who followed common practices, which is why he saved the best wine for last.

They say the best wine is the one you like the most. Perhaps that's true, but the most expensive wine is the aged one.

Wine is a product of maturity. The wine Jesus produced was more valuable than the wine served at the beginning. Jesus not only ensured that the party didn't run out of wine but also accelerated the maturation process to make the wine of the highest quality.

Grape juice is like a child. Young wine is the one that goes straight from the steel tanks to the bottles, meaning it goes from the fermentation columns to being sold. This is the cheapest wine. Then there is the aging wine. It goes from steel to oak and is allowed to ferment for some time to mature. If it ferments for three years, it is called Reserva, and if it ferments for five years or more, it is called Gran Reserva - always written in uppercase letters.

Our character needs to transition from being grape juice to Gran Reserva. Our ministries need to transition from being grape juice to Gran Reserva. Our technology programs need to transition from being grape juice to Gran Reserva.

The Most Perfect

Now I invite you to think theologically. I want to apply the maturity model to the maturity of God. I acknowledge that this is a novel idea, but I invite you to continue reading and judge it at the end of the chapter.

I must clarify that we cannot confine God within the framework of a human model. In other words, God does not conform to our models. However, is not it the task of systematic theology to create models of thought? Isn't theology a method for organizing our knowledge of God? Remember that if something is systematic or methodical, it is technology, according to the definition in **Chapter 3**.

Reflecting on the maturity of God considering the maturity map from the previous chapter certainly does more for us than for God. First, it allows us to discern some attributes of who God is, and second, it validates the maturity map considering that revelation of who God is in the Bible. So, let us begin.

The Lord Jesus Christ exhorted us to be perfect:

Matthew 5:48: "Be perfect, therefore, as your heavenly Father is perfect."

How is this possible? One of the most foolish statements anyone can make is, "I am not perfect." Why? Because we think it is obvious that no one is perfect on the face of the earth. Even trying to be perfect is a futile task. No matter how hard anyone strives, they can never become perfect.

I do not know where the idea comes from that only God is the one who is perfect and that no one else can be, but it does not come from the Bible. Thinking that way would invalidate the words of Jesus. According to Jesus, it is indeed possible to be perfect, just as the heavenly Father is perfect. So, what is perfection all about?

The Greek word that is translated as "perfect" in this passage corresponds to the word *téleios*, which means complete, fully developed, accomplished, or lacking nothing. Biblical perfection is maturity.

Biblical perfection is maturity.

A more accurate translation of this passage would replace the word "perfect" with the word "mature." In fact, it makes more sense in the context of the discussion where Jesus is talking about loving those who are not so lovable.

The call of Jesus is to be mature, just as God the Father is mature; and mature in love.

This is a fundamental premise:

In the Bible, God is the most mature being in the universe. He is the Great I AM – and you always write it in capital letters.

Perfection is not an unattainable level of impeccable and absolute morality, as some commentators assume. Perfection is maturity and the fulfillment of your purposes. It is the highest degree of fruitfulness. The words of Jesus give us an impetus to strive for perfection, to strive for maturity. It is possible.

The Ancient of Days

Maturity can also be understood in terms of age, for example, when we say that a person is mature in years. The book of the prophet Daniel describes God as someone who is advanced in age.

The New International Version translates it as the "Ancient of Days" (Daniel 7:9, 13, 22), while the New Living Translation refers to Him as the "Ancient One" (Daniel 7:9, 13, 22). The idea is of someone who has reached the fullness of days and who will be present "in the latter days."

This does not mean that God has gone through literal growth like any other human being. "Ancient of Days" is not a title but a description, so it is not something that was achieved but something that is inherent. Daniel uses this expression as a figure or a metaphor to describe a God who is venerable, a King who is worthy and wise, and a Judge who will sit on the throne to restore the kingdom to His people.

> *God is the most mature being in the universe.*

The word that is translated as "ancient" in these passages is the Hebrew word *attiq*, which means aged or old; it has gone through the maturity that time brings. Hebrew is a language of action, which means that all words are derived from a verb. The corresponding verb to *attiq* is *atheq*, which is translated as weaning or being detached.

Weaning is the end of breastfeeding, the second cutting of the umbilical cord. It marks the transition from infancy to childhood. It means that the child will no longer depend on the mother's body for nourishment and, in terms of food, can self-sustain.

Both the apostle Paul and the apostle Peter used weaning as a symbol of spiritual maturity or its delay, as immaturity. Paul writes:

> Brothers and sisters, I could not address you as people who live by the Spirit but as people who are still worldly—mere infants in Christ. I gave you milk,

> not solid food, for you were not yet ready for it. Indeed, you are still not ready. You are still worldly. For since there is jealousy and quarreling among you, are you not worldly? Are you not acting like mere humans? (1 Corinthians 3:1-3).

Peter writes in en 1 Peter 2:2: "Like newborn babies, crave pure spiritual milk, so that by it you may grow up in your salvation."

Of course, there is no maturity without growth. The author of the book of Hebrews agrees with this image when he writes:

> In fact, though by this time you ought to be teachers, you need someone to teach you the elementary truths of God's word all over again. You need milk, not solid food! Anyone who lives on milk, being still an infant, is not acquainted with the teaching about righteousness (Hebrews 5:12-13).

When we speak of the Ancient of Days, we infer someone who has been weaned long ago. The Ancient of Days is an expert in the message of justice. He is the Master par excellence, the Savior who sealed salvation. He does not behave according to human standards, for He is not a mere mortal who lies (Numbers 23:19). He is the only truly self-sufficient being.

The Most Fruitful

Generally, the act of weaning focuses on the one being weaned and not on the mother. However, what happens with the mother is relevant to our analysis of maturity.

The hormone of lactation, called prolactin, suppresses the hormone of gestation, called progesterone. The effect is that during lactation, the possibility of pregnancy is reduced. These values are reversed with weaning: prolactin decreases, and progesterone consequently increases. God's creation is perfect! Weaning allows the mother to become pregnant again. Weaning makes the mother more fruitful and able to reproduce.

There is a hidden lesson here regarding our technology initiatives. Innovation is an act of letting go of the old, breaking away from the familiar. To achieve technological maturity, we need to wean ourselves from our childish programs.

As I demonstrated in **Chapter 1**, maturity is related to the ability to bear fruit. The more mature, the more fruitful, and vice versa. So, if God is the most mature being in the universe, He should also be the most fruitful.

Well, I have news for you: God is the most fruitful being in the universe.

God is the origin of all things (Genesis 1). "Through him all things were made; without him nothing was made that has been made" (John 1:3).

God is the one who endowed living beings with the ability to grow and reproduce. God created us in His image and likeness. He made us prolific beings. This capacity to multiply applies not only in Genesis, as we have already studied, but in both paradise and captivity, in agricultural and urban societies.

God said to His people when they were in captivity in Babylon:

> Build houses and settle down; plant gardens and eat what they produce. Marry and have sons and daughters; find wives for your sons and give your daughters in marriage, so that they too may have sons and daughters. Increase in number there; do not decrease. Also, seek the peace and prosperity of the city to which I have carried you into exile. Pray to the Lord for it, because if it prospers, you too will prosper (Jeremiah 29:5-7).

God is the one who poured out the Holy Spirit of fruitfulness for the birth and reproduction of the church. The church multiplied through the outpouring of God's Spirit and continues to grow through the same Spirit.

God is the most fruitful being in the universe.

Maturity is related to being fully united. Without divisions. We have all received a map of unity and perfect integration in Jesus Christ, as the apostle Paul states:

Ephesians 4:13: "Until we all reach unity in the faith and in the knowledge of the Son of God and become mature, attaining to the whole measure of the fullness of Christ."

The fruit of the Christian is the fruit of the Spirit, which enables us to become mature Christians.

Jesus Christ is the model of maturity, and He gives us the measure of the stature we should attain. He and the Father are one (John 10:30). He prayed for us to be one as well (John 17:22).

Furthermore, Jesus cried out on the cross, saying "it is finished," that His purpose was fully accomplished, and His work was complete (John 19:30).

The author of Hebrews reminds us that He is the author and perfecter –the one who brings our faith to maturity:

> Fixing our eyes on Jesus, the pioneer and perfecter of faith. For the joy set before him he endured the cross, scorning its shame, and sat down at the right hand of the throne of God (Hebrews 12:2).

A perfect technology is not one that has no errors, bugs, or failures. Perfect technology is one that does everything it needs to do, fulfills the purpose for which it was created, and is integrated at a high level.

Perfect technology is mature technology.

For example, a chair is perfect when it does not lack a leg and when you can sit on it with confidence, even if the paint is chipped or it does not match the other furniture in the room.

The Most Mature

Let us remember for a moment the maturity roadmap:

Maturity roadmap:

Initiate
Inquire
Investigate
Implement
Ideate
Identify
Innovate

Certainly, we cannot confine God within any model; He surpasses every model. However, let us now explore God in the light of the maturity map to discover how mature He is.

The purpose of examining God within the maturity map is not to validate God Himself, but rather to validate the map itself, ensuring that it is not only technologically consistent but also theologically sound. Let us proceed.

Initiate

God initiated everything out of nothing. This is known in theology as *creatio ex nihilo*. He created everything. He knows everything. He understands everything. He does not need to learn anything. The prophet expressed this with a series of rhetorical questions.:

Isaiah 40:14: "Whom did the Lord consult to enlighten him, and who taught him the right way? Who was it that taught him knowledge, or showed him the path of understanding?"

The answer is no one. God did not need consultants to enlighten Him. God had no teachers in matters of justice. No one imparted knowledge to God. There is no being smarter, less ignorant than Him. There is no being more capable of initiating something than Him.

This stage does not apply to God.

God is omniscient. There is no ignorance in Him, nor has there ever been. God's omniscience eliminates ignorance.

Inquire

God is love. He does not wander inquiring or being curious. He is not inclined to investigate or learn because He knows everything. There is no area of knowledge that is beyond His reach.

God is not timid. There is no fear in Him. He has not given us a spirit of timidity either (2 Timothy 1:7). Maturity leads us towards the absence of fear. As we draw closer to our model in God, He frees us from fears.

Psalm 34:4: "I sought the Lord, and he answered me; he delivered me from all my fears."

God is love, and in Him, there is no fear, "but perfect love drives out fear" (1 John 4:18). Once again, the word "perfect" appears, which we should understand as "mature." Mature love is, as I explained earlier, perfect love.

For God, this stage does not apply.

God is love (1 John 4:8), and there is no intrigue in Him. God's perfect love eliminates all intrigue and uncertainty.

Investigate

God can do all things. He does not need to try anything. He does not need to experiment. He can do it perfectly from the start, and it will turn out well. God does not test things to see how they turn out and then make corrections. He does not throw things out to see what sticks.

God is not like Microsoft, which, when it released Windows 11, had hundreds of known imperfections and was considering feedback from beta users to determine which ones to fix.

God can do whatever He purposes in His heart to do, and He is perfect from the beginning; He is flawless. He does not need to ask what needs correction. He does not need beta users.

Furthermore, God knows the intentions and attempts of humankind long before they result in success or failure.

Jeremiah 17:10: "I, the Lord, search the heart and examine the mind." God is the only one who can truly examine the hearts of men, not the other way around.

For God, this stage does not apply.

God is omnipotent. There is nothing He cannot do. The omnipotence of God eliminates the need for investigation, trial, and error.

Implement

Dios is the ultimate planner.

Studying God's actions in history can indeed lead to the development of a theology of planning, as suggested by Tony Morgan in his book *Developing a Theology of Planning*. He writes:

> There is an overwhelming amount of Scripture that supports planning. And I think most leaders would agree that vision must be translated into an action plan. Otherwise, the best of intentions will become the regrets of tomorrow. [20]

When it comes to creating a plan and implementing it, God is the star.

His plan was so effective that it became one of the most powerful and transformative theological truths in human history: it is known as the Plan of Salvation.

In our time, He will reveal the plans He has for us.

Jeremiah 29:11: "'For I know the plans I have for you,'" declares the Lord, 'plans to prosper you and not to harm you, plans to give you hope and a future'."

As God knows all things in advance (foreknowledge) and is not subject to time (eternity), we can say that God has already created all His plans and has successfully implemented them. We need to mature into our future. God is not subject to the variables of time and space.

He continues to save. He continues to write the plan and the story of salvation in the hearts of men, winning them one by one.

This stage does not apply to God.

God is omnipresent. There is no place or time in human experience where He is not present and fulfilling His purposes. Although we may not be able to see it, God's plans are made and executed. The omnipresence of God eliminates the stage of implementation.

God has already created all His plans and has successfully implemented them.

Ideate

God is the great thinker. His thoughts are deeper, more abundant, and higher. They are perfect. They are mature.

In the face of God's profound ideas, the psalmist exclaims, "How great are your works, Lord, how profound your thoughts!" (Psalm 92:5). The passage suggests that God's works are a direct result of His thoughts. God acts in complete harmony with what He thinks. His works are great because His thoughts are deep.

Considering the abundance of God's ideas, the psalmist continues to exclaim, "How precious to me are your thoughts, God! How vast is the sum of them!" (Psalm 139:17).

Regarding the quality of God's ideas, the prophet writes, "'For my thoughts are not your thoughts, neither are your ways my ways,' declares the Lord. 'As the heavens are higher than the earth, so are my ways higher than your ways and my thoughts than your thoughts'" (Isaiah 55:8-9).

God is an expert in all matters. He has brought His thoughts to the highest level of maturity. He is the reference point. The counselor. In fact, one of God's titles is the Wonderful Counselor, as referred to in Isaiah 9:6:

> For to us a child is born, to us a son is given, and the government will be on his shoulders. And he will be

> called Wonderful Counselor, Mighty God, Everlasting Father, Prince of Peace.

As a counselor, God has the best advice; He produces the best ideas. Earlier, I mentioned Isaiah 40:14, but the preceding verse, verse thirteen, reads: "Who can fathom the Spirit of the Lord, or instruct the Lord as his counselor?"

The answer once again is no one. In fact, we know the Spirit of the Lord or the Holy Spirit as the other Comforter or Counselor (John 14:16, 26; 15:26).

This stage does not apply to God.

God is the Wonderful Counselor. God's counsel surpasses the stage of ideation because all His thoughts find concrete expressions in timeless reality; all His counsel is eternal. God is, and He knows who He is.

When Moses asked God by whose name he should go to the people, God revealed His name.

Exodus 3:14: "God said to Moses, 'I AM WHO I AM. This is what you are to say to the Israelites: I AM has sent me to you.'"

God does not have issues of identity or dissociation. He is one. He is one within Himself, He is one with the Son, and He is one with the Spirit of unity.

Deuteronomy 6:4: "Hear, O Israel: The Lord our God, the Lord is one."

Jesus Christ said, "I and the Father are one" (John 10:30).

God's people's unity should reflect this intrinsic unity of God. "Make every effort to keep the unity of the Spirit through the bond of peace" (Ephesians 4:3). The unity of the Spirit is a characteristic of a mature church.

Furthermore, God distinguishes Himself above all others. God is not only one, but also unique.

Identify

We can translate the same passage in Deuteronomy 6:4 as: "The Lord our God is the only Lord." There is no other like Him, nor will there ever be.

For God, this stage does not apply.

God is unique. There is nothing that compares to Him. The uniqueness of God places Him beyond identification.

Innovate

Finally, God is God of the new. Innovation is the level of maturity promised by God through the words of the prophet:

Isaiah 43:19: "See, I am doing a new thing! Now it springs up; do you not perceive it? I am making a way in the wilderness and streams in the wasteland."

Once again, He is opening a path to pass through and rivers to renew us. This is the image: When unexplored terrain opens to the world with the first steps of the traveler, it is God who goes ahead, cutting through the vegetation. When a small idea makes its way into a sea of possibilities, it is God who initiates the brainstorming.

Isn't "brainstorming" one of the most successful and widely used creative techniques?

When it says "once again," could it be because God has done it before and still has new things up His sleeve? –if God wears robes with sleeves.

Appropriately, the last book of the Bible speaks of seven new things that God is working on. As they are new, they are not yet in sight.

God is God of the new.

Seven is a symbolic number. It does not mean that God is working on only seven new things, but rather that God's innovations are complete and perfect.

Revelation 21:5 reads: "He who was seated on the throne said, 'I am making everything new!' Then he said, 'Write this down, for these words are trustworthy and true'."

These seven things encompass "all things":

1. A new heaven.
2. A new earth.
3. A new city.
4. A new relationship.
5. A new temple.
6. A new light.
7. A new paradise.

By the way, if in physics light was the beginning of everything we know (Genesis 1:3) and establishes the measure of time to which we are subject in this dimension, it is fitting to think that new light will give rise to a new dimension of life and a redefinition of space/time as we know it.

Innovation is the highest level of maturity, and the recreation of "all things" is the highest achievable level of innovation.

Some may think that God is not doing anything new because we do not see it, but God is creating many new things that are beyond our limited capacity to see.

If we were to ask what God is doing now, we would have to say that He is working on all the new things that are yet to come.

God is an innovator. He dwells in His eternal innovation.

Let us review:

God knows everything.
God is love.
God can do anything.
God is the Ultimate Planner.
God is the Great Thinker.
God is unique.
God is innovative.

This allows me to conclude that the maturity map presented in the previous chapter has a profound theological significance. It is valid and useful in the exercise of ministry mediated by technology.

CYBERMATURITY

You have reached the highest level of maturity when you can have peace amid the storm.

– JIM REEVE, AMERICAN PASTOR.

Fourteen days of continuous anguish. The shadow of death covered all the crew members. Euroclydon struck the ship with all its furious hurricane force, while Poseidon did everything in his power to sink it with his fierce trident. There was no rest whatsoever. In their desperate struggle to survive and remove the giant waves that filled the deck, they had even forgotten to eat, which worsened things because no one realized their own weakness. They became increasingly ineffective without knowing it.

It is curious. There were experienced sailors and a trained captain, there were men of war and a centurion, and yet, you could not find the necessary leadership to calm the spirits. So much experience together but so little expertise. There was no one who acted with composure and a clear mind, let alone somebody that could think of the correct strategy. In fact, they were in those circumstances because they had ignored the warnings and plotted the navigation map incorrectly. The wrong map can get you into serious trouble and lead you to an undesired destination.

Only one of the prisoners raised his hand, his voice, and took the initiative. No sane person would think of celebrating the Lord's Supper during such turmoil, except for him. One who was a true leader. It is because a true leader knows how to listen to the Master of the winds and storms. He may not have walked on the waters of the Sea of Galilee, but now he had his own opportunity to see Jesus resting in the boat. That is why he could have peace amid the storm. I am talking about the apostle Paul.

He had opposed continuing the journey, but he all ignored him, and now they were in serious trouble that threatened the lives of the two hundred seventy-six people on board, frightened, weary, and weak. During the meal, he comforted them by saying that God would preserve the lives of all, but not the ship or its cargo. Thanks to providence, the shipwreck occurred when the ship ran aground on a sandbar near an island, and everyone was able to swim toward it.

Acts twenty-seven narrates the episode in detail. I would like you to take a couple of minutes to read the entire chapter before continuing.

Storms in Ministry

How did Paul end up in this situation? The narrative begins in Acts 21. He had the intention of going to Rome, but the opportunity had not presented itself. On his last visit to Jerusalem, while preaching in the temple, some Jews from Asia who were familiar with his ministry stirred up the crowd and incited them against him. As the mob gathered to attack him, they alerted the commander of the Roman battalion, who ordered his arrest and flogging.

While in prison, Paul invoked his Roman citizenship to avoid the lashes, and a series of legal hearings ensued, starting with the Council of Jewish authorities. As a Roman citizen, the local authorities transferred his case to the governor of Caesarea, Felix, who delayed the legal process for two years until his successor, Festus, arrived. The new governor consulted the case with King Agrippa, grandson of Herod the Great, before whom Paul appealed to Emperor Nero. He was then put on a ship bound for Italy.

The journey across the Mediterranean from the coastal town of Caesarea in the region of Sharon to Rome was quite challenging due to logistical issues and the atmospheric changes signaling the approach of winter.

Luke, the narrator of Acts, accompanied Paul on his journey and described the route from departure to shipwreck. In fact, there were 7 stops of immaturity leading to disastrous results: Caesarea, Sidon, Cyprus, Myra, Cnidus, Fair Havens, and Malta. The only fair thing about the journey was the name of the port before the shipwreck.

Note the vocabulary of the story. The winds were contrary (v. 4), the navigation was slow, they barely reached the next port, the wind was unfavorable to continue the original course (v. 7), but they continued with difficulty (v. 8). They wasted a lot of time, and the navigation was dangerous (v. 9). What seemed appealing turned into a nightmare (v. 13). They were adrift (v. 15), lost their bearings, and eventually all hope of getting saved (v. 20).

This could very well describe your experience in ministry in general and specifically the use of technology for ministry purposes. You work in constant opposition, feeling like you are not making progress, moving only laterally and with great difficulty. Everything seems to conspire against you, and you find yourself going in a direction that unintentionally takes you away from your original purpose. Though you manage to continue, you waste a lot of time on distractions and dangers. You drift aimlessly.

Certain opportunities appear attractive but continue to lead you astray. One link leads you to another and another, and you end up on a page you did not want to see. You are carried away by circumstances, events, and what the congregation decides. You lose your bearings. You lose strength, and even the last bit of hope.

That is how storms in ministry are. But remember, Jesus is still resting in your boat. You can still listen to Him, and even when you consider everything is lost, the most important thing is you.

The first part of your experience using technology in ministry may have ended in a shipwreck. So, for the second part of your journey, chart a roadmap to maturity that can lead you to your destination. Remember that Paul was able to eventually reach Rome.

Cybernauts

You may be wondering at this point, what does this story have to do with assessing maturity? Well, it has a lot to do with it. Luke narrates that when Paul told the centurion not to continue the journey, "But the centurion, instead of listening to what Paul said, followed the advice of the pilot and of the owner of the ship" (Acts 27:11).

The Greek word for pilot or captain of a ship is *kybérnetes*. It is a compound word that means to guide or steer navigation or *nautikós*, also from Greek.

In ancient times, the pilot not only steered the ship but also managed everything related to the vessel. The Greek historian Plutarch writes in this regard: "The pilot selects his sailors, and the owner selects the pilot."[21] Early on, the term began to be used metaphorically for life, the ship, which needs to be guided by a captain. But who is the true captain of the ship?

Plato attributes the piloting of life to God.[22] Philo of Alexandria refers to Moses as a man who submitted himself to the direction of divine reason, [23] and the narrative of the martyrdom of Polycarp XIX:2 says that Jesus is the Pilot of our bodies. [24]

Paul knew this well. Although he had no voice or vote, his life was subject not to the centurion, the captain, or the owner, but to the will of God, who had promised him that he would reach his destination. Furthermore, if Paul was not the most experienced sailor in the waters of the Mediterranean, he was the most experienced among shipwreck survivors (2 Corinthians 11:25).

The duties of the pilot included planning the route, making decisions along the way, reading navigation instruments, selecting his crew of sailors, and overseeing the operation of the ship. These management actions were known as *kybérnesis*, often translated as administration.

Paul himself identifies with a particular group of ministers appointed by God who are entrusted with the administration of the church. He writes in 1 Corinthians 12:28:

> And God has placed in the church first of all apostles, second prophets, third teachers, then miracles, then gifts of healing, of helping, *of guidance*, and of different kinds of tongues (*italics* added for emphasis).

Those who guide (*kybérnesis*) are part of the leadership that God designed for the church. The idea of piloting the church may have a precedent in the book of 4 Maccabees 7:1, which, although not a canonical book, is likely that Paul was familiar with it, and it says: "For like a most skillful pilot, the reason of our father Eleazar steered the ship of religion over the sea of the emotions." [25]

Cybernauts are those called to lead the ministry. They are the leaders who understand that Christ is the owner of the ship. They navigate the route of maturity. They choose their team, chart, and follow the maturity roadmap from port to port, recognize times of change, and know how to read the navigation instruments to make wise decisions and avoid shipwrecks.

Cybernetics

The word "cybernetics" derives from this background. That is why we do not ride or fly through the Internet, but rather navigate.

Think for a moment about ancient navigation. It is easy to notice a few things. The rudder controlled the direction. The speed depended on the sails and, in some cases, the number of rowers. The location was determined by reading the position of the sun and the stars.

Luke is precise in indicating that for many days "neither sun nor stars appeared" (Acts 27:20), and that is why they lost all hope of being saved from the storm. They had lost their instruments. They were disoriented and adrift. They could not know where in the vast sea they were or which way they were heading.

Not knowing their position and direction, they feared the ship would strike some unforeseen rocky formation and send them to the bottom of the sea with no chance of survival.

Instrumentation with all its measurements and calculations is critical for cybernetics. Instrumentation allows you to know where you are –by the position of the stars, where you are heading –by the position and trajectory of the sun, and how far you have yet to go –by the speed of the winds.

Merriam-Webster defines cybernetics as "the science of communication and control theory that is concerned especially with the comparative study of automatic control systems." [26]

Cybernetics has been strongly associated with computers, interchangeably. The reason is that computers are the most sophisticated automation and communication machines of human ingenuity. The interaction between humans and computers has developed our capacity in extraordinary ways.

What facilitates this identification between computing and cybernetics is the ability of computers to provide control, measurement, and automation. In fact, computing has changed navigation forever. Satellite-based navigation systems such as GPS (Global Positioning System), autopilot systems, and various sensors enable the control of even the most complex vessels. We can ensure a high degree of safety in navigation thanks to the instruments that we employ.

> *Cybernetics applies to a vessel and a computer with the same effectiveness that it applies to an organization like the church.*

Cybernetics is not limited to computers. It encompasses other exchanges between the living and the non-living, between humans and technology. Cybernetics applies to a vessel and a computer with the same effectiveness that it applies to an organization like the church.

Today's church needs cybernauts. Where are the cyberministers of today? A good cybernaut will understand the technological resources available to the ministry and will use them to avoid organizational shipwrecks. They will be able to read the cultural climate and chart a safe course. They will enable the church or institution to save its cargo, but more importantly, all its crew members. The mission of the cybernaut is salvation, both personal and organizational.

On the contrary, a bad cybernaut ignores the navigational conditions, the characteristics of the vessel, and the suggestions of the crew. The only guarantee a bad cybernaut can offer an organization is shipwreck.

I want to illustrate this with a real-life case. The ministry of God is Love church set out to implement a membership and ecclesiastical management system. These systems consist of a database and software application with forms to interact with the database. The database stores personal information of the members along with the history of contact between the church and each person. Additionally, it could create groups of people using a variety of criteria for Bible study and common interests.

The church leadership decided to purchase the system without conducting a cost analysis, without comparing the features of this system with others of similar purpose, and, worse yet, without evaluating the supplying company.

First mistake: They made a blind purchase, without a point of reference. Never start a journey without knowing where you are. This should have been defined in the first place (Step #1: Initiate).

Once they closed the contract, the software company sent a specialist to implement the system, make necessary adaptations to align it with the church's ministry practices, and train the administrative assistant.

Second mistake: They started the implementation process (Step #4) and customization (Step #6) without thoroughly studying the organizational needs at all levels (Step #2). You cannot skip the steps of maturity. It was certain that this system would affect not only the administrative assistant but also the overall functioning of the church in each of the five areas of technology, as we will see later.

When the specialists completed their work, they left without proper documentation. After a few weeks, the church realized that the administrative assistant could not handle all the tasks alone and that the training she received was insufficient for her to train others.

Third mistake: There was no proper testing and adjustment period (Step #3) that was long enough to understand the impact on the church's regular operations and all its employees and volunteers.

After one year, and amidst many difficulties the church faced in using the software, the vendor announced that they would discontinue support for that product. Now the church was worse off than before, frustrated by the time wasted and all the money spent on a discontinued product.

Who should have anticipated this situation? Who could have prevented the losses and frustrations? Was it the specialist? The administrative assistant? The software vendor? Certainly not. The person responsible was the cybernaut, the cyberminister.

A good cybernaut understands technological maturity and uses a navigation roadmap. They do not skip any of the steps. On the other hand, a bad cybernaut ignores maturity; due to lack of knowledge or lack of interest, they lead technology to failure.

In **Chapter 5**, I discussed the maturity roadmap that allows you to define the location and direction of your technological initiatives. However, my experience tells me that many ministries already have some form of location and direction, but they lack measurement. Do not think you are alone in this. Many technology departments in recognized institutions across the country suffer from the same problem.

When we apply cybernetics to the maturity of ministerial technologies, we arrive at a three-component model: location, direction, and measurement. Notice the parallel with the three-dimensional purpose of technology.

As a result, we may have a vague idea of progress, but we do not know for certain if we are maturing. We do not have the means to confirm or assess it. We lack the proper navigation instruments.

We need a way to evaluate maturity. A way to measure it. Do you want to know if your ministerial technology programs are yielding the desired results? You must evaluate them. Do you want to know if ministerial technology is maturing? You must evaluate it over a frequent and extended period.

Cybermaturity

The purpose of measuring maturity is not to obtain absolute results that determine the value of your initiative, but rather to be able to compare those results over time with your own previous measurements. In other words, it is not about knowing whether your ministerial technological initiative is good or not, but about determining if it is growing or maturing.

Measuring technological maturity allows you to identify trends that will inform your growth.

Measuring technological maturity allows you to identify trends that will inform your growth.

Fran Buytendijk, in his book *Performance Leadership*, explains how measuring accomplishes the following three objectives:

1. Being accountable to comply and report to the governing members.
2. Facilitating strategic decision-making and learning for leaders (cybernauts).
3. Guiding people's behavior. [27]

People. Your team. The members of your technology team. The members of your congregation. Everyone can align their activities with the goals of technological maturity as you can demonstrate progress. People will abandon dysfunctional behaviors and sabotage when you make better decisions and learn in the process.

I do not know if this is a Eureka moment for you, but it certainly was for me. If you can demonstrate through evaluation that you are progressing or maturing, you will gain the trust of those who follow you.

And thus, we arrive at defining cybermaturity:

> Cybermaturity is the ability of leaders to define the location, direction, and measurement of technology usage in ministry in route towards innovation.

I have defined technology in **Chapter 3** as the integration of five elements: people, systems, processes, policies, and devices. Well, the maturity of any technological initiative should be measured in each of these five elements of technology. Technology should not only mature, but it should do so by balancing each of these five areas.

Let us see what this means in the following examples:

People

Cybermaturity applied to your technology personnel involves evaluating their technical skills, creating an update program, and assessing not only their learning but also their ability to apply the acquired skills to ministry practice. Individuals seek growth on the path to cybermaturity.

For example, let us consider Cornelius, who is interested in assisting with audiovisual media. What do you do? You need to determine his existing skills, identify areas for improvement, and provide training to acquire those skills. Cornelius must undergo a maturity process that allows him to embrace the mission, understand the organizational culture, adhere to established processes, and handle the equipment. He needs technical and organizational training.

For Cornelius, there are specific initiation activities, and as he matures in his role, he can gradually reach levels of ideation, identification, and even innovation. Education should permeate every maturity process. You should not leave to chance or circumstances your staff technical education; you must plan it and map it out.

Keep in mind that technical education consists of acquiring knowledge and immediately applying it in practice. If there is a considerable time gap between the two, the acquired knowledge may dissipate and lose its effectiveness.

Today, there are many viable alternatives for informal education that can help you acquire the variety of technical skills you need to perform your job. Training is a shared responsibility. Your ministry should create scheduled spaces for technical education. Cornelius also needs to maintain a learning mindset. That is why I often tell my employees, "If you don't take two to three hours each week to learn and stay updated, you are not doing your job well."

Systems

Cybermaturity applied to your systems involves evaluating their effectiveness, undergoing a reengineering process, and assessing whether the system changes increase effectiveness or not. Systems aim for effectiveness on the path to cybermaturity.

For example, let us consider the Sunday worship organization system. Currently, everything depends on the pastor: choosing worship songs, collecting announcements for the week, preparing the sermon, contacting all participants, and assigning responsibilities.

This system is not remarkably effective, not because the pastor is incapable of handling everything, but because it revolves around a single person. What if the pastor is unavailable, falls ill, or goes on vacation? Additionally, it is not a sustainable long-term model as it dilutes the pastor's focus between routine activities, cuts on pastoral care, and reduces his or her time for sermon preparation.

A more mature model would be to assign a worship director. The worship director would be responsible for coordinating trivial activities while the pastor focuses on ministerial duties. This model aligns more with biblical principles. [28]

Indeed, knowing how to delegate is a sign of maturity. While the pastor may have to do everything in the beginning, as they delegate tasks, they can reach higher levels of maturity and even innovate.

Knowing how to delegate is a sign of maturity.

Processes

The cybermaturity applied to your technology processes will allow you to evaluate them critically and adjust them to be repeatable and consistently reproducible in the shortest possible time.

Processes aim for efficiency on the path to cybermaturity.

For example, let us consider the process for collecting and counting tithes and offerings. I have been in churches where prayer is offered during the collection of offerings, and ushers pass containers to receive them. I have also been to other churches where attendees are pressed for half an hour before prayer and ushers are allowed to pass.

Both are effective (offerings are collected), but which process is more efficient? Which process is more sustainable? Another way to think about it is in terms of economics: which process is more cost-effective? In the case of this example, which process allows for better offerings? I do not know. It may depend on the culture of your church.

Cybermaturity helps you evaluate processes to make that decision. You can use one method for three months and measure the results, then change the method and measure again. You will have a comparative reference, and based on that, you can determine your response.

You could then implement a modern system for tithe collection through text donations, allowing for recurring and automatic giving using debit cards, credit cards, or electronic checks, instead of passing collection bags during the service. In any case, you must evaluate the results.

Policies

Cybermaturity applied to your technology policies aims to better protect your assets, the security of your systems, the replication of your processes (fruitfulness), and the contributions of your members.

Policies strive for normalcy on the path to cybermaturity.

For example, consider the policies governing the operation of your website. Your website should include a copyright statement to protect your intellectual property; it should ensure technical security through encryption, password standards, and access restriction policies; it should standardize the programming and publishing of content by creating templates and standards; and it should have visible usage policies and adhere to them.

All these policies become even more critical if you receive online donations or have an online shopping store, or if you collect personal data during event registration processes (remember FERPA from **Chapter 3**). If your website operates in the United States, it must comply with the regulations of the Federal Trade Commission (FTC).

It is not about being "legalistic" but being "protective."

Few ministries take on the task of creating these policies. This is one of the most deficient areas in the Christian organizations I have had the privilege to work with. If that is your case, you need to initiate a process of cybermaturity right away.

Devices

Cybermaturity applied to the devices you use in your ministry aims to protect your investment in terms of their function.

Devices seek support and maintenance on the path to cybermaturity.

For example, the lighting system in the sanctuary may include controls to regulate natural light filtered through skylights, high-powered lamps for the auditorium, specialized spotlights for the stage, flashlights for musicians, and emergency lights. All of this requires support and maintenance.

You must provide support to the electrical system, lighting control consoles, and emergency lights. You need to maintain cables, bulbs, batteries, or fuses by managing the devices with care and replacing each component at the end of its lifespan.

A mature technology program provides a budget for receiving technical support when needed, maintaining devices in a timely manner, and replacing them when they expire. It is essential to have a technology replacement plan that gradually updates the devices and provides replacement options in case of unexpected failures.

To recap, on the path to cybermaturity:

- *People strive for growth.* A more mature technology staff has more knowledge and is better prepared.
- *Systems strive for effectiveness.* A more mature technological system achieves its goals more frequently and with greater assurance.
- *Processes strive for efficiency.* A more mature technological process consistently reproduces the same results with a higher level of accuracy and reliability.
- *Policies aim for normalcy.* A more mature technology policy provides greater security and better controls risks.

- *Devices aim for support and maintenance.* More mature devices are less prone to unexpected failures.

Benefits

Cybermaturity offers a multitude of benefits. In **Chapter 3**, I described the three purposes of technology: strategy, tactics, and intelligence. Cybermaturity offers benefits in each of these areas.

Cybermaturity helps you determine your current state, set your direction, and implement controls for evaluation. It integrates the three purposes of technology as previously defined. Location aligns with strategy, direction aligns with tactics, and evaluation aligns with intelligence.

Cybermaturity aligns with your strategic goals, tactical objectives, and intelligent decisions.

Cybermaturity will save you thousands of dollars and countless headaches.

Cybermaturity will foster loyalty and result in long-term alliances with your employees, volunteers, and service providers.

Cybermaturity will help you recognize negative trends and apply timely corrections to reverse them.

Cybermaturity will provide you with simple yet powerful evaluation tools.

So, how do we evaluate technology in ministry? How do we measure maturity in concrete terms? This is the topic of the next chapter.

ASSESSING MATURITY

Maturity is that age in which you stop lying to yourself.

– RALPH WALDO EMERSON, AMERICAN ESSAYIST.

Imagine the perplexity of a man outside time and space, who has lost his watch, and his measuring rod, and his tuning fork. I believe, Sir, that it is indeed the state which constitutes death.[29]

– ALFRED JARRY, FRENCH SYMBOLIST WRITER.

A couple of years ago, I had the privilege of implementing a disaster recovery system for a certain educational institution. The project aimed to protect financial information in a geographically distant location so that, in the event of a natural disaster in the area, the systems could continue functioning in the other location within minutes.

When high availability systems like these are established, there are two measurements that regulate the technical characteristics of the design and implementation. These are known as MTBF (Mean Time Between Failures) and MTTR (Mean Time to Recover).

The goal is to have a much higher MTBF, calculated in months or even years, while keeping the MTTR as low as possible, measured in minutes or seconds. In other words, the aim is to have longer intervals between system failures and a much faster recovery time.

I recall that before we even began the implementation, one of the systems experienced a failure. The main hard disk containing the database became corrupted and stopped working. It was the third time that year it had happened, but since the disk operations were not being monitored, the administrators did not realize it and could not move the information to the backup disks before the failure occurred.

Fortunately, we had configured the system to take nightly backups and store them in a location separate from the damaged computer. However, we did not have spare hard disks of the same model required by the computer. We had to order the disks from the manufacturer, but they did not have them readily available because those computers were discontinued, so a special order had to be placed. The disk arrived at the office a week later, and thanks to the configured backup, we were able to recover the system.

These experiences allowed us to calculate an availability baseline that looked like this:

MTBF = 4 months
MTTR = 10 days

As you can imagine, valuable hours of work were lost during the week and a half that the system was unavailable. Entire departments were paralyzed. When the system returned to normal, the institution had to hire temporary employees to catch up with data entry and the resulting backlog.

MTBF and MTTR have a direct impact on the finances and operations of any organization. In an immature system, failures occur more frequently, and the time to recover is much longer.

The disaster recovery project took these factors into account. The goal became to configure a system that extended MTBF to several years and reduced MTTR to less than ten minutes. We implemented a process of technological maturity that, I am pleased to report, was successful.

How do I know this? Because we conducted several tests simulating system failure and consistently activated the recovery process in less than ten minutes. We measured the results each time and kept a log with relevant information to confirm our progress.

If something is worth it,
it is worth measuring it.

If something is worth it, it is worth measuring. If you lose your watch, your ruler, and your tuning fork, you are in trouble; if you lose your measuring instruments, you can consider it lost. But the more detailed your accounting, the faster your recovery will be. That is why we need to evaluate our maturity processes.

Furthermore, continuous evaluation keeps you accountable to others and grounded so that you do not deceive yourself. Want to know if church attendance is growing or declining? You must measure it. Want to know if people are visiting your websites? You need to use measurement tools like *Google Analytics*. Want to ensure that people are watching the worship service through your mobile app? Then the platform you use should include ways to measure participation details.

Teamwork

In this chapter, I want to suggest three evaluation tools: the *Attitude Scale*, the *Principles Lanes*, and the *Technology Wheel.* However, before I do that, I want to ensure that these tools are used as part of a team evaluation.

As you will see later, your team is critical to any technological development. This also applies to evaluating maturity. Depending on the size of your ministry, organize a group of people who can assess the maturity of technology. A team of 5 to 10 members is ideal.

It is preferable that each of these individuals is at least familiar with the concepts in this book, although it would be much more effective if they read it together as a group. Here are some ideas to initiate this preparation process would be:

1. Purchase a copy of this book for each team member and set a deadline for reading.
2. Divide the team into small groups of no more than 3-4 people for reading and discussion.
3. Create a book club on Goodreads. Goodreads is a website that allows you to keep track of all your books and create book clubs. This way, you can see all the team's comments in one place.
4. Direct team members to join the official Cybermaturity book club on Goodreads. This way, you can participate in the conversation alongside other ministries going through the same process. All books cited in the bibliography are included in this group for easy reference: (www.goodreads.com/group/show/1214350-cybermaturity).
5. Discuss the questions in **Appendix D: Discussion** at the end of the book together.
6. Organize a pre-evaluation meeting where you review the concepts of the maturity map.

7. Although the focus of this book is not personal application, ask each evaluation team member to apply the concepts to their use of a specific technology.

Decide in advance how you will initiate the process, following some of these ideas. Once you decide on the method you will use, proceed to organize your evaluation team.

Your team is critical to any technological development.

The composition of your team is crucial. I suggest including a congregation member who is not involved in leadership to capture the perspective of the members as well. Your technology maturity evaluation team in the church would be composed as follows:

1. *The pastor*: The pastor's participation is critical. Along with the ministry leaders, the pastor is responsible for defining the strategic direction of the organization.

2. *Ministry area leaders*: Similarly, the evaluation of technology maturity has strategic purposes, and it is crucial for leaders to participate in the assessment. The specific leaders will depend on your church's or institution's particular organization structure.

3. *The technology and media directors*: These may not be official titles, but you may have individuals responsible for coordinating sound, audiovisuals, or office systems.

4. *Members of your technology team*: Delegated by the technology director to perform certain routine functions. They may be

the most technologically skilled individuals in the church or organization.

5. *Lay member*: Although I dislike the distinction between laypeople and clergy, as we are all priests to God, I understand the distinction. This is a member of the congregation who identifies as an end user of technology.

Ensure an elevated level of commitment to the process. Team members should meet deadlines for reading the book, for example, attend regular meetings if you chose to form small groups, or write their personal evaluations to share with the rest of the team.

Each member should respond with brutally honest feedback, to the best of their knowledge and with a willingness to learn. When your team is ready to conduct the evaluation, use the three tools provided below.

Attitudes Scale

This is the question we need to answer first: What is the rate of immaturity in our attitudes towards ministry technologies? We do this by evaluating attitudes towards maturity.

Do you remember the three areas of cybermaturity? By accounting for attitudes towards maturity in our technology programs, we create a visual map of location. Even better, if we track maturity over an extended period, we can identify the direction of trends to determine if we are progressing or drifting.

While you could evaluate technology in the church, this assessment will make more sense if you place it in the context of a specific technology initiative. For example, what are the attitudes towards the church's website? Or what are the attitudes towards the membership system?

Name of the team member:										
Fear										Love
-5	-4	-3	-2	-1	0	1	2	3	4	5
Resistance										Meekness
-5	-4	-3	-2	-1	0	1	2	3	4	5
Conformity										Entrepreneurship
-5	-4	-3	-2	-1	0	1	2	3	4	5
Blindness										Vision
-5	-4	-3	-2	-1	0	1	2	3	4	5
Comfort										Innovation
-5	-4	-3	-2	-1	0	1	2	3	4	5
Sterility										Fruitfulness
-5	-4	-3	-2	-1	0	1	2	3	4	5
Grand total										

Figure 3: Attitudes Scale

The scale shown in **Figure 3** represents the evaluation of attitudes towards technological maturity in ministry.

The scale ranges from negative attitudes to replacement activities, using the terminology from **Chapter 2**: from graffiti to the blackboard. Assign values from -5 to 5 to the perception of each attitude, and then calculate a final overall attitude score. This final score reflects the general sentiment of the evaluating team.

To use the attitude scale, follow these simple instructions:

1. Review **Chapter 2** if you need to refresh the concepts about attitudes of immaturity.
2. Commit each team member to be honest in their responses (we are truthful Christians).

3. Have each team member circle their perception of each of the six represented attitudes. You can use **Appendix A** for this purpose.
4. Connect the circles with a line, which will provide a visual representation of your current state.
5. Calculate the total points for each section and the overall total for the scale.
6. Create a small table to record the results of each team member and calculate the final totals.
7. Calculate the averages for each area and fill in a new attitude scale with these averages.
8. Save the results for future reference.

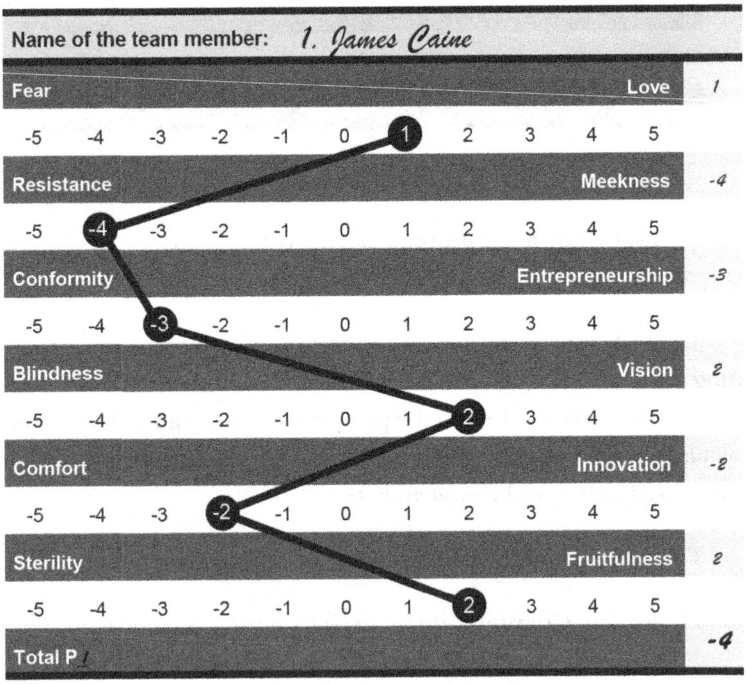

Figure 4: Attitudes Scale Example

At the end of the evaluation, the attitude scale will look as shown in **Figure 4**. The ideal state is when all the circles align with the same value on the positive side of the scale. This rarely happens.

Each team member will have a different perception from their own point of view, and that is okay. Some members may have a more positive evaluation, while others may have a more negative one, and that is also acceptable.

It is likely that optimists and pessimists will emerge. Validate each team member for their participation and do not judge their responses.

Remember, the purpose of the exercise is to determine the overall sentiment regarding the maturity of technology in your ministry.

Name	Fear	Resistance	Conformity	Blindness	Comfort	Sterility	Grand Total
Total							
Averages (Total/)							

Figure 5: Attitudes Scale Totals by Team

You can see a straightforward way to calculate the totals per team in **Figure 5**. Write the names of each participant and their totals. Then calculate the overall totals for each attitude and the averages. Finally, calculate the grand total and grand average.

When completing the information, the table will look as shown in **Figure 6**.

The maximum points that the most optimistic member can give is thirty (30) and the most pessimistic is minus thirty (-30), which corresponds to the maximum score (5) for each of the six (6) attitudes.

Name	Fear	Resistance	Conformity	Blindness	Comfort	Sterility	Grand Total
James Caine	1	-4	-3	2	-2	2	-4
Patricia Johanson	2	-1	-2	2	0	1	2
Mary Pratt	2	-1	-3	1	0	1	0
Peter Nichols	2	1	3	0	1	2	9
Nathan Smith	-3	-1	1	2	2	2	3
Total	4	-6	-4	7	1	8	10
Averages (Total/ 5)	0.8	-1.2	-0.8	1.4	0.2	1.6	2

Figure 6: Attitudes Scales Totals by Team Example

Calculate the total of points the whole team could score.

Maximum by team = 30 x (___ # participants)

Minimum by team = -30 x (___ # participants)

The minimum and maximum values establish the range of the scale per team. Calculate the difference between the grand total and the team scale and calculate the percentage of points obtained relative to the total possible points. This percentage represents the overall sentiment of the team regarding the maturity of technology in your organization.

In the example from **Figure 6**, the team scale ranges from -150 to 150. This is a range of three hundred (300) points. The team's grand total is ten (10). This would correspond to an overall sentiment expressed as a percentage of (160/300), which is 53%. This indicates a slightly favorable sentiment with ample room for growth.

Graph the overall averages on a new scale (like **Figure 4**). This scale represents the values of the entire team. Once you complete the evaluation, proceed to analyze the results. Here are some questions that can facilitate the discussion about the *Attitude Scale*:

1. What is the significance of the overall sentiment percentage?
2. Does the overall sentiment align with the team's subjective perception?
3. What can we do to remedy the overall sentiment?
4. Who is the most pessimistic team member and why? And who is the most optimistic?
5. Which attitude received the lowest and highest scores?
6. What activities can we promote to improve negative attitudes?
7. How does this evaluation compare to previous evaluations? What are the trends?
8. Once you have completed the three tools, return to the *Attitude Scale,* and ask: How do the data from the *Attitude Scale* inform the *Principles Lanes* and the *Technology Wheel*?

Please let me know if there is anything else I can assist you with.

Principles Lanes

The purpose of this tool is to answer the question: How much do the principles of technological maturity govern our ministerial practice? Leading by principles is like flying by instruments.

By accounting for how much the five maturity principles govern our technology programs over an extended period, we create a visual map of direction. You can think of the five principles as lanes on a racetrack, and they are as shown in **Figure 7**.

Maturity aims to change human behavior in their interaction with the environment and technology. I am not an expert in psychology, but behavior is modified by replacing undesirable behavior with desired behavior over a sustained period. You do not wake up one day at the next level of maturity suddenly, but through small gradual changes that start from nothing.

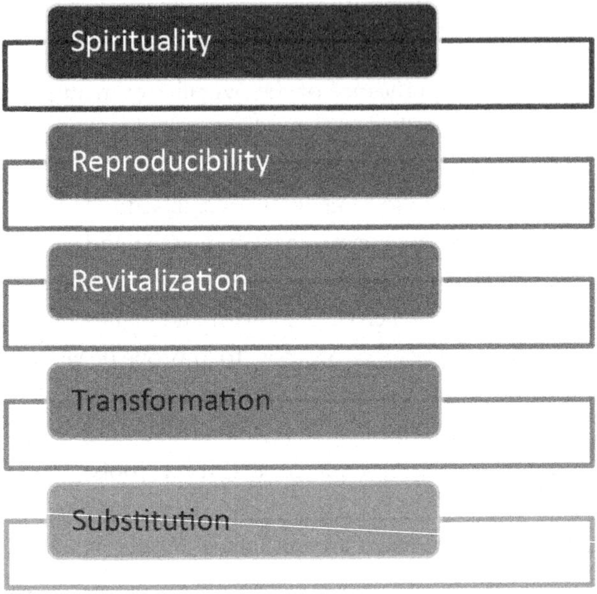

Figure 7: Principles Lanes

If your technology program is consistently progressing over time, it is maturing, but if it is declining, it is not. It is that simple. A cold spirituality ceases to reproduce; if it stops reproducing, it loses vitality; without vitality, there is no transformation, and without visible transformation, desirable behaviors start being replaced by undesirable ones.

That is why the feeling of regression that you sometimes experience is more than just a feeling. It is an accurate perception of reality. If you are not moving forward, you are moving backward. Even stagnation, which could be a valid stance, is considered regression when you consider it in relation to the context. Isn't it critical then to measure the direction of maturity?

Figure 8 provides a sample questionnaire for evaluating maturity principles. You will find the complete questionnaire in **Appendix B**. If you would like to obtain an electronic copy of the questionnaire and all the other evaluation tools for printing purposes, visit www.cyberministry.info.

#	What is your level of agreement with these statements?	None	Low	Some	High
1	We have been able to discern our calling to ministry and are taking steps to excercise it.	☐	☐	☐	☐
2	If I think about the ability of technology to prolong life, I can cite some examples in my mind.	☐	☐	☐	☐
3	The vision of the church is written in detail and the leaders understand were we are heading as an organization.	☐	☐	☐	☐
4	I aspire with confidence for the church ministry to progress to the next level of technological maturity.	☐	☐	☐	☐
5	The church seeks to explore different alternatives to solve the technological problems it faces on a daily basis.	☐	☐	☐	☐
6	Technology helps us live in devotion to God, knowing that our primary goal is to worship Him.	☐	☐	☐	☐
7	Everything we do and everything we should do must be guided by excellence; excellence glorifies God and adds more honor to His name.	☐	☐	☐	☐
8	There is a direct relationship between the technology I use and the growth I see in the ministry; better technology leads to greater growth.	☐	☐	☐	☐
9	Every time there is a change, it is easy for me to adjust my behavior and routine to adapt.	☐	☐	☐	☐
10	We have defined a roadmap that everyone can follow to become a disciple of Jesus Christ, and we actively follow it.	☐	☐	☐	☐

Figure 8: Questionnaire: Principles Lanes of Technology Maturity

The evaluation of maturity principles is done quantitatively using this standardized questionnaire, which consists of thirty-six (36) statements and a four-point scale rating from 0 to 3.

0 = None
1 = Low
2 = Some
3 = High

There are six (6) statements for each principle with six (6) related questions regarding strategic mission matters. Additionally, they are not organized in any apparent order.

To fill out the questionnaire, follow these simple instructions:

1. Review **Chapter 4** if you need a refresher on the concepts expressed about maturity principles.
2. Encourage each team member to be honest in their responses. Remind them that the idea is to capture their personal appreciation of technology. Honesty will not be penalized.
3. Let each team member mark their response to each statement. Encourage them not to overthink their answer but to mark the first response that comes to mind.
4. Create a table with the assigned values of 0-3 for each participant's responses.
5. Group the scores for each of the five principles for each participant. Follow the "Questionnaire Answer Key" in **Appendix B** to do this.
6. Create a small table to record the results for each member and calculate the final totals for the entire team.
7. Write the values in your *Principles Lanes.*
8. Save the results for future reference.

The totals for the entire team can be recorded in a table like the previous section, but the values will be much higher due to the questionnaire collecting the data.

Upon completing the information, the table will appear as shown in **Figure 10**.

The maximum number of points that the most optimistic member can assign for each principle is eighteen (18), which corresponds to the maximum score of 3 for each of the six (6) questions assigned to each principle, resulting in a maximum total of ninety (90) points for the entire principles questionnaire.

Name	Substitution	Transformation	Revitalization	Reproducibility	Spirituality	Grand Total
Total						
Averages (Total/)						

Figure 9: Principles Lanes Totals by Team

It does not include the eighteen points that would be assigned to the strategic questions. The lowest possible value is zero (0). Note that there are no negative values.

Calculate the maximum total points possible for the entire team. In our example with five (5) participants, it would be four hundred fifty (450) points.

Maximum by team = 90 x (___ # participants)

The range of the team's scale is determined by the minimum and maximum values. Calculate the difference between the overall total and the team's scale range and calculate the percentage of points obtained compared to the total possible points. This percentage represents the overall direction of the team in terms of the technological maturity principles in your organization.

In the example shown in **Figure 10**, the range of the team's scale is from 0 to 540. The overall total for the team is 230. This would result in a general predisposition towards maturity expressed as a percentage (230/450) of 51.1%. The dominant principle would be transformation.

Name	Substitution	Transformation	Revitalization	Reproducibility	Spirituality	Grand Total
James Caine	11	15	9	4	6	45
Patricia Johnson	8	16	6	1	10	41
Mary Pratt	9	13	5	4	10	41
Peter Nichols	11	14	7	11	9	52
Nathan Smith	12	16	8	7	8	51
Total	51	74	35	27	43	230
Averages (Total/ 5)	10.2	14.8	7	5.4	8.6	46.0

Figure 10: Principles Lanes Totals by Team Example

Record the overall averages in lanes (like those in **Figure 7**). These lanes represent the values of the entire team.

Now, calculate the group's perception regarding the ministry strategy with the corresponding questions. There are six (6) questions with a maximum score of three (3) points each. In our example with five (5) team members, this would represent a maximum of 15 points for each question. Fill in a table with the team's totals as shown in **Figure 11**.

Strategic Area	
Vision	
Mission	
Values	
Strategies	
Discipleship	
Impact	

Figure 11: Strategic Areas Table

Many other data and calculations can be derived from all the collected data using this evaluation tool, but the most important aspect is to use it to facilitate discussion among your team. Here are some questions that can help facilitate the discussion about the *Principles Lanes*:

1. What is the significance of the overall predisposition percentage?
2. Does the overall direction align with the team's subjective perception?
3. What can we do to redirect the course if necessary?
4. Who is the most pessimistic team member and why? And who is the most optimistic?
5. Which principle obtained the lowest scores? And which one obtained the highest scores?
6. What activities do we need to replace to move towards spirituality?
7. How does this evaluation compare to previous evaluations? What are the trends?

8. Once you have completed all three tools, go back to the *Principles Lanes,* and ask: How do the data from the *Principles Lanes* inform the *Attitude Scale* and the *Technology Wheel*?

Engaging in these discussions will provide valuable insights and help guide your team towards maturity in technology within your organization.

Technology Wheel

Finally, you need to answer the question: What is the account balance of each of the technology areas in our ministerial practice? That is the purpose of this tool.

By accounting for each of the five technology areas with respect to the seven steps, we complete the measurement of cybermaturity. Let us recall that the five areas, based on my definition of technology from **Chapter 3**, are as follows:

1. People
2. Systems
3. Processes
4. Policies
5. Devices

And the levels of maturity (**Figure 2**) are:

1. Initiate
2. Inquire
3. Investigate
4. Implement
5. Ideate
6. Identify
7. Innovate

Figure 12 below represents this.

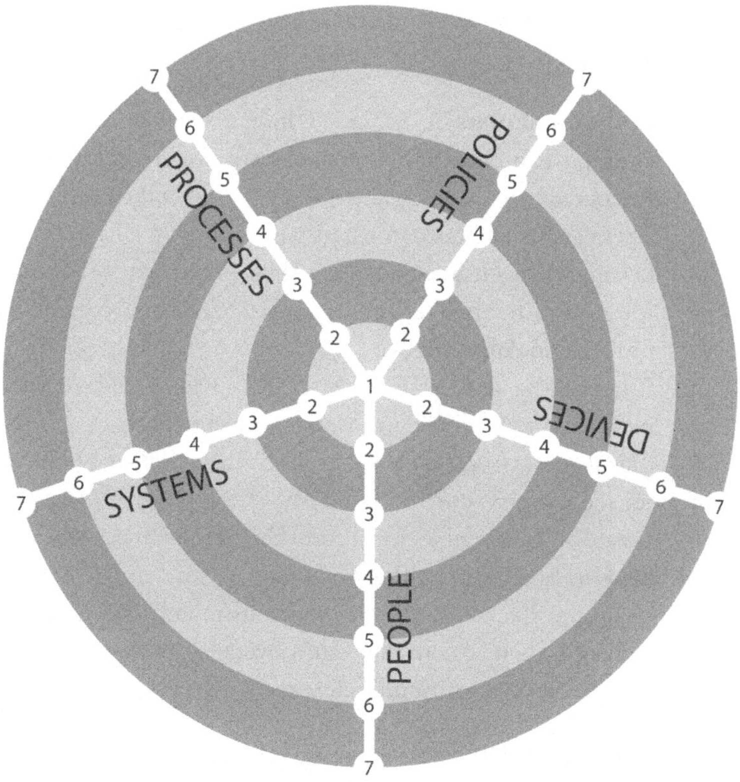

Figure 12: Technology Wheel

The purpose of this tool is to assess the maturity level of each technology area in your organization and identify areas for improvement. By evaluating where each area falls within the maturity levels, you can determine the overall balance and progress of your organization's technology practices.

The maturity of each technology area can be evaluated using a wheel of concentric circles. The numbers 1 to 7 represent the steps or stages of maturity, and the concentric circles correspond to these steps. The five technology areas are represented by the five radial lines.

To assess the state of technology in your ministry, follow these instructions:

1. Review **Chapters 3**, **5**, and **7** if you need to refresh the concepts expressed about technology and the maturity map.
2. Encourage each team member to be honest in their responses. There are no right or wrong answers.
3. Have each team member mark their perception of maturity for each of the five technology areas. You can use the wheel provided in **Appendix C**.
4. Connect the dots with a curve. This will provide a visual representation of the current state of technology.
5. Calculate the total points for each section and the overall total for the wheel.
6. Create a small table to record the results for each team member and calculate the final totals.
7. Calculate the averages for each area and fill in a new technology wheel with the team's overall averages.
8. Save the results for future reference.

The figure you will obtain will be something like **Figure 13**.

In the example, you considered that you have a maturity level of 4 for the devices you have purchased but a level of 3 for the people needed to manage them. This represents an imbalance. The same process can be applied for each of the five technology areas, assigning a maturity level from 1 to 7.

Perfect maturity would imply a perfect wheel with a level 7 for each area, which is not impossible but difficult to achieve. It represents the ideal state.

If your technology wheel is far from being a perfect wheel, be prepared for a bumpy ride. The wheel is a graphical representation of the state of technology in your ministry and identifies the areas that need to grow.

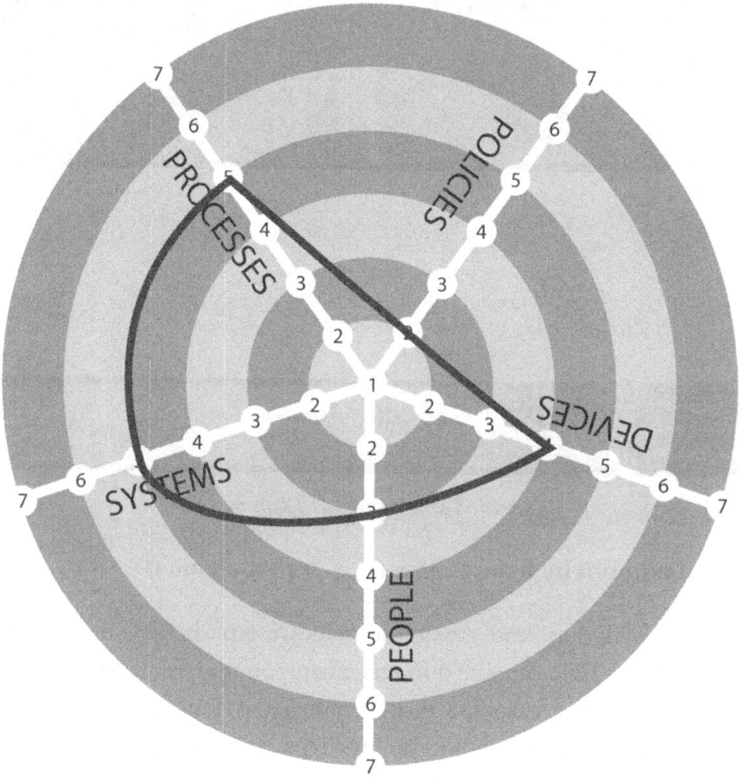

Figure 13: Technology Wheel Example

The totals for the entire team can be recorded in a table like the one in the previous section as shown in **Figure 14**. And upon completing the information, the table will look exactly as shown in Figure 15.

The maximum number of points that the most optimistic team member can assign when summing up the five areas is thirty-five (35), which corresponds to the maximum score of seven (7) for each of the five areas (5). The most pessimistic score is zero (0). Note that there are no negative values.

Calculate the maximum total points possible for the entire team.

Name	People	Systems	Processes	Policies	Devices	Grand Total
Total						
Averages (Total/)						

Figure 14: Technology Wheel Totals by Team

Maximum by team = 35 x (___ # participants)

The minimum and maximum values establish the range of the team's wheel. Calculate the difference between the overall total and the team's range and calculate the percentage of total points obtained relative to the total possible points. This percentage represents the stage of cybermaturity in which your organization is located, if you place each cyber maturity step on a scale that increases by 14% (100% divided by seven steps).

In the example from **Figure 15**, the range of the team's scale goes from 0 to 175 (5 members x 35 points each). The overall total of the team is seventy-five (75). This would correspond to a technology maturity level expressed as a percentage (75/175) of 43%, which would place your team at Step 3 or "Investigating".

Record the overall averages on a new technology wheel (like **Figure 12**). This wheel represents the values of the entire team.

Name	People	Systems	Processes	Policies	Devices	Grand Total
James Caine	3	5	2	1	4	15
Patricia Johnson	3	3	1	2	3	12
Mary Pratt	4	5	3	1	3	16
Peter Nichols	4	4	2	1	3	14
Nathan Smith	3	4	3	3	5	18
Total	17	21	11	8	18	75
Averages (Total/ 5)	3.4	4.2	2.2	1.6	3.6	15

Figure 15: Technology Wheel Totals by Team Example

Finally, discuss the meaning of these results with all members of your team. The following questions can facilitate the discussion about the *Technology Wheel*:

1. What does the level of cybermaturity signify?
2. Does the overall maturity align with the team's subjective perception?
3. What can you do to balance the wheel at the highest level reached in the most mature areas?
4. Who is the most pessimistic team member and why? And who is the most optimistic?
5. Which technology area obtained the lowest and highest results?
6. What factors can drive you towards innovation?
7. How does this evaluation compare to previous assessments? What are the trends?
8. Now that you have completed the three tools, return to the *Attitudes Scale* and *Principles Lanes,* and ask: How do the data from the *Technology Wheel* inform the *Attitude Scale* and the *Principles Lanes*?
9. What conclusions can we draw?

10. Take note of all evaluation conclusions and save the document for future reference.
11. Create a list of actions you will take and set deadlines for each action. For each action, specify which evaluation result you intend to address.
12. Assign team members responsible for following up on each growth alternative.

You can use this *Technology Wheel* to assess the overall state of technology in your organization, but you can also use it to evaluate specific technologies on which you are focusing maturity efforts. This means you will create multiple Technology Wheels during your assessment, such as one for musical instruments, one for audiovisual equipment, one for membership software, one for the offering system, and so on.

Recommendations

The evaluation of technology maturity in ministry is a process that involves a high level of subjectivity. It is based on individual perceptions of attitudes and how well the evaluation team believes the church or organization aligns with certain general principles. I am aware of this.

Some may argue that subjectivity cannot be measured. However, if this were true, nearly all research methods in the sciences would be a waste of time.

William Thomson, the Irish physicist, and mathematician, once said:

> When you can measure what you are speaking about and express it in numbers you know something about it; but when you cannot measure it, when you cannot express it in numbers, your knowledge is of a meagre and unsatisfactory kind. [30]

Who was William Thomson? He was none other than Lord Kelvin, the creator of the Kelvin temperature scale (°K). In his time, it was believed that absolute temperature could not be measured because it was considered a subjective concept.

By assigning values to concepts that are relatively subjective, we can analyze trends in these values and the overall sentiment of the participants. Numbers, even if they are simple, are the beginning of knowledge.

This is no small matter. If the leadership of your organization "feels" that there is progress, it is possible that there is, and this feeling can empower them to continue working. But if they do not feel the progress, they will be demotivated and hinder growth.

Numbers, even if they are simple, are the beginning of knowledge.

An example of this is using technology to communicate the vision of the church. This would mean that the vision is communicated not only from the pulpit but also visually through printed banners, audiovisual materials, and your digital presence.

You can produce videos that reinforce the vision. You can compose music and incorporate it into your worship service with the theme of the vision for the congregation to sing. You can even create a digital media kit for members to share your vision on social media.

This could be a good alternative use of ministerial technology; it would not only involve the church in technology-related matters but also potentially yield more visible results as people grasp the vision. Other methods for measuring vision can be devised using the three tools suggested in this section.

On the other hand, it is possible that sentiment does not fully align with reality when analyzing these trends. One way to counteract this is by forming a diverse evaluation team, as described in a previous section of this chapter. An honest, sincere, and committed team can make a difference.

Another way is to hire an external advisor who can guide the process and provide a more objective perspective from the outside. I will discuss this topic in **Chapter 10**.

I would like to make some final recommendations regarding the evaluation process:

1. Conduct an initial evaluation before embarking on a technology initiative. This initial evaluation establishes a baseline for future comparisons.

2. Perform periodic evaluations. It's a good idea to conduct evaluations on a quarterly, biannual, or annual basis. Decide on the frequency of your evaluations in advance.

3. Encourage your entire leadership team to fill out the evaluations independently, and then calculate the totals and averages to obtain a final evaluation.

4. Compare your results with the baseline and previous periods to determine trends.

5. Determine whether there is progress or regression on your path to maturity and correct the course if necessary.

Every measurement tool should be taken in the overall context of your ministry and leadership team. In this regard, I leave you with the words of William Cameron, who wrote:

> Not everything that can be counted counts, and not everything that counts can be counted. [31]

The next chapter applies the maturity concepts discussed so far to a specific example of technology in ministry to connect them to practical application.

APPLYING MATURITY

To exist is to change, to change is to mature, to mature is to go creating oneself endlessly.[32]

– HENRI BERGSON, FRENCH PHILOSOPHER.

We arrived at my friend's house with great expectations. The gathering was rather simple. The atmosphere was festive yet relaxed. We had been wanting to get together for months, but our respective busy schedules had gotten in the way. As we arrived one by one, we settled comfortably in the living room with our sandwiches from a nearby deli and something to drink.

We waited for several minutes, engaging in casual conversation in anticipation. My friend is a successful businessman who has long desired to support Christian ministries with a good reputation and track record. He had instilled in us the same desire to serve, which is why he had gathered us together.

After a while, the guest of honor arrived. If you could call him that, he is a well-known Christian celebrity in Latin America. He was accompanied by his right-hand man in the ministry, an engineer who had left his career to join him and provide corporate representation. We talked for hours. He gave us an overview of his ministry work and brought samples of his sermons. His approach was something like this:

"This is my ministry work. I want to continue doing the same and have you help me distribute it to a larger number of people."

I do not want to discount or minimize his ministry work in any way. He has done more for Latin American families than many politicians, economists, or businesspeople could ever achieve. His work is exceptional, and its impact is significant. That was not the problem.

The problem with our guest of honor was a matter of maturity. It wasn't that he was an immature person, quite the opposite. It was about the technological maturity of his ministry. He wanted to keep doing the same without seeing the need to recreate or reinvent. He was also not willing to multiply himself in others.

Although his teachings are powerful, some of his materials are still distributed on magnetic tapes or compact discs. The recordings are of very low quality - you can even hear people's shouts and dogs barking in the background. His digital presence is very poor, and his digital identity is scattered.

Our guest needed to undergo a process of reengineering. To mature, you need to constantly recreate the forms of ministry. The ministry itself remains the same, but the form must be subject to scrutiny and constant change.

Reverse Engineering

If you have ever played with building blocks, you know that to create something new, you must dismantle what you had previously built. To be able to build, you must first deconstruct. One common form of deconstruction (and therefore construction) is reverse engineering.

In reverse engineering, you start with the end in mind. You break down all the parts, disassemble all the blocks, and document the details of each step with the aim of being able to reproduce the model.

Reverse engineering is appropriate for reducing the complexity of an object or system, discovering its flaws, and facilitating reuse. It serves as a mechanism to envision your new state of maturity and facilitate innovation.

Innovation does not mean that something is entirely new and that no one had thought of it before. It means that it is new to you or your organization. It is highly likely that what you want to achieve, or build has already been accomplished by someone else, and if not, there is certainly a similar model that you can use as a foundation.

To mature, you need to constantly recreate the forms of ministry.

In that case, innovation involves taking the precedents you know and adapting them to your specific circumstances. That is why in every innovation, there is an element of continuity and an element of breaking away from what already exists.

You apply reverse engineering to the continuity aspect to introduce contextual changes that break with the original design and create something new. Remember the words we have already studied from Isaiah 43:19: "See, I am doing a new thing!" means using what already existed to create something new.

The idea I suggest with this is that you can start by analyzing the components you observe in a ministry that is more technologically mature than yours, sift them through the maturity map, and apply reverse engineering principles to your own ministry or organization.

This process can be seen in **Figure 16** on the next page.

Digital Presence

In the remainder of this chapter, I want to deconstruct the digital presence of a hypothetical ministry called Omega to apply the insights to your own ministry, assuming that Omega is a good model to follow for what you want to achieve.

In my book, *Cyberpresence*, I define the concept of digital presence with all its components and provide you with a practical guide to establish a solid digital presence for your ministry.

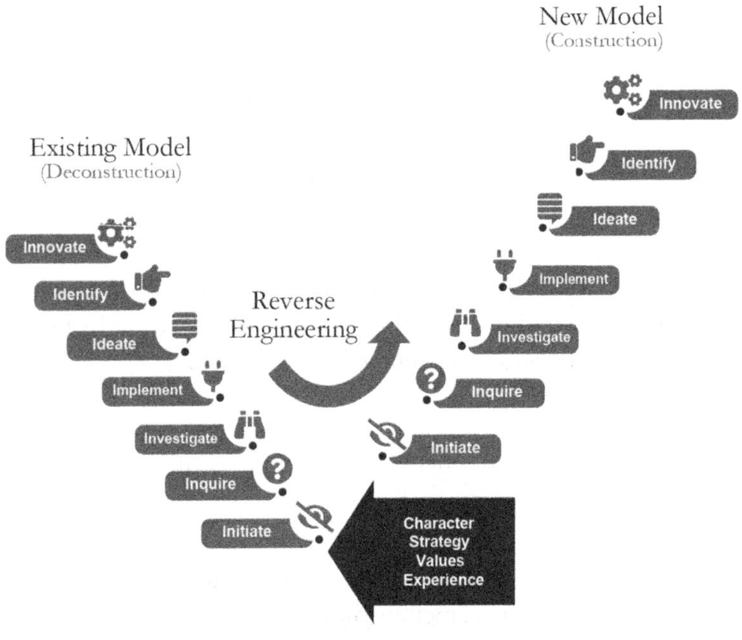

Figure 16: Reverse Engineering

For this, I will apply the maturity map (**Chapter 5**) in reverse, observing each of the five elements of technology (people, systems, processes, policies, and devices), as defined in **Chapter 3**.

Omega's digital presence is exemplary. It is comprehensive, integrated, and clear. They have developed the potential of their ideas, their website, the use of email, and their presence on social media.[33]

Their outreach strategy has yielded good results, reaching hundreds of thousands of followers with their message.

Omega has a multiplier emphasis, not only in their process of gaining followers but also because they have replicated themselves in hundreds of similar ministries. They even organize an annual convention for independent churches that follow their model to update them on the use of cutting-edge technologies.

They effectively utilize multiple online tools and maintain a platform that encourages contributions from many other Christian leaders. With their participation policies and information security measures, they have earned the trust of the community. Their ability to create high-quality digital products helps local ministries in their respective missions. Through their self-sustaining management, they generate financial resources to support other organizations.

The employees and volunteers are committed to Omega's vision and carry out the corporate mission. They are trained and qualified, highly productive, and have opportunities for personal and ministerial development.

In every innovation, there is an element of continuity and an element of breaking away from what already exists.

Now, let us engage in step-by-step reengineering. We need to ask ourselves: how did Omega look in each of the technology areas at each stage?

Let us see.

Cybermaturity 7

If we evaluate the maturity of Omega using the tools from **Chapter 8**, we will know that it is at the *Innovate* stage, for the following reasons:

People: There is diversity. There is strategic alignment with the vision and mission. They have the necessary skills to perform their work. They are creative and contribute ideas. They have a sense of purpose.

Systems: They generate finances and have good administration. They follow a clear and integrated model of digital presence. They reach their community and develop their leaders. It is evident that their training programs are effective.

Processes: They understand the production cycle and follow quality control procedures. They have documented the flow of tasks and the expected outcomes for each method. They have done the same for contributors and those who wish to replicate their model.

Policies: They have defined qualifications for employees, volunteers, content contributors, and those accessing their resources. They uphold privacy, confidentiality, and information rights.

Devices: They have a stable and always available platform. They understand that different tools and software applications are needed for each type of work. They have plans for support, maintenance, and device renewal.

Cybermaturity 6

It corresponds to the *Identify* stage. There is an emphasis on assimilation. How did Omega look in each of the technology areas when it was at Cybermaturity 6?

People: Personnel were reviewed to ensure all necessary employees were in place, and spaces were created for volunteers. People were trained in their respective areas, both in understanding the corporate culture and possessing technical skills. The purpose was for individuals to adopt the corporate identity of the ministry and recognize technology as essential for mission fulfillment.

Systems: Administrative systems for personnel, finance, community outreach, e-commerce, and learning were integrated to share information and operate collectively. Many standards were documented, and data duplication was minimized as much as possible.

Processes: Routine tasks were simplified and automated. Production cycles were defined, and quality control procedures were established. Staff identified with their role in the workflow. Significant corporate documentation was generated. Manuals were created for collaborators and replicators.

Policies: The directive was established that technology would become a strategic advantage and that progress would not be possible without it permeating every area of the ministry. Policies on the use of digital resources, confidentiality, copyright, and others were written and published. Job profiles for employees and descriptions of volunteer opportunities were documented.

Devices: More sophisticated and advanced devices were acquired. A device renewal plan was implemented, supported by the overall budget. A central monitoring system with alerts and notifications was put in place to ensure device availability and performance, as well as their respective services.

Cybermaturity 5

This level corresponds to the *Ideate* stage or thinking about a future that does not yet exist. There is an emphasis on creativity. Let us see how Omega looked in each of the technology areas at this Cybermaturity stage 5:

People: The leadership created the corporate identity of the ministry through a participatory strategic planning process. Clear goals were established, and it was decided to embed technology within the mission. The entire team was aligned with the vision and mission of the ministry, including the technical staff.

Systems: Regular brainstorming sessions, feedback loops for creativity, and ongoing evaluation meetings were established. Production and content validation systems were created, along with a regular update calendar. A social media management system was implemented.

Processes: The foundation for process analysis was established, leading to the creation of initial flowcharts, and areas for possible automation were identified. Clear lines of responsibility were established for executing specific tasks.

Policies: The need for clear policies in areas such as information security, protection of personal identity, intellectual property, and data backup was recognized. Initial documents were drafted with an informed legal basis. Software purchase contracts were reviewed to find optimal and cost-effective long-term solutions.

Devices: A thorough review was conducted to determine if the existing devices were suitable. New productivity and creativity tools were acquired, and staff were trained to make better use of these tools.

Cybermaturity 4

This level corresponds to the *Implement* stage and has an emphasis on routine work. What activities characterized Omega in each of the technology areas during this stage of cybermaturity?

People: The achievement-oriented focus of this stage led Omega to seek qualified personnel for tasks and hire consultants to complement the workforce. The staff were trained in the lacking technical skills. During this time, they hired a Director of Ministerial Technologies and integrated them into the ministry's organizational structure.

Systems: Decisions were made regarding what to do and what not to do. The main systems for digital presence, content production, and all supporting systems such as payroll, finance, donations, personnel management, and membership were implemented. Project management practices were established.

Processes: The organization's processes were adjusted to assimilate the new systems. Each final product began to be viewed as the result of a process, and consistency patterns were adopted, even if they were not well-documented yet. Clear ownership of each step in the processes was defined.

Policies: The legal implications of relevant matters were discovered. For example, issues related to corporate identity or tax payment to the national treasury. Legal advice was obtained. Special attention was given to software licenses and equipment warranties.

Devices: Lights, camera, action. The first professional equipment was purchased, including computers and networks, hosting and cloud services, and auxiliary software to facilitate task execution.

Cybermaturity 3

This is the level that corresponds to the *Investigate* stage, focusing on exploring and experimenting. It was the time of the early adventurers, those who felt the need to try new forms and means of communication. What activities characterized Omega in each of the technology areas during this stage?

People: Some young individuals started creating small video snippets, and the pastor began writing a devotional to send via email to close contacts, just to see if it would catch on. Another adventurer started recording the pastor's messages, without a clear perspective of where it was heading, but at least to have a record.

Systems: They were not the most advanced or sophisticated. In fact, they used only free or inexpensive tools because it was not worth paying for something they were unsure would yield results or that they wanted to pursue in the medium or long term. The sound quality of the recordings was not excellent, but at least they had them.

Processes: They did not give much thought to processes. They did what they could with what they had and whoever was available at the time. There was not a detailed plan, but there were plans to do something different. The trial-and-error process was organic and intuitive. They did not have quality standards.

Policies: There was a set of basic rules set by the pastor or director, but there was still a lot of flexibility in following them. They did not think about the legal implications of their attempts. They did not pay attention to the fine print of licenses.

Devices: They seemed more like hobbyist devices. It was not necessary to buy sophisticated or professional equipment. If it did the job, even if the quality was poor, it was welcome. Some devices were received as second-hand donations, but they helped in their experimentation.

Cybermaturity 2

This is the level that corresponds to the *Inquire* stage. It's the time of curiosity and the death of the cat. How did each of Omega's technology areas look during this stage? Not much, not clearly; rather like a glimpse of something nascent.

People: Some were interested in exploring a digital presence. They saw something on television or started following another slightly more advanced ministry. The source does not really matter. What is important is that this small interest turned into intrigue and led them to investigate, seek resources, and compare opinions. No one was formally in charge, but some showed some interest. Someone came up with the idea that we could do something similar.

Systems: It was common to say, 'We had nothing.' But that was the reality. Much was done through word of mouth. The only existing system was research, and it was not even done formally. There was no maturity in the research.

Processes: Nothing formal or centralized. Everyone inquired on their own, and occasionally, in a casual way, they shared information about their findings with others. They gathered common interest. No one ventured to try anything during this time.

Policies: What is that, and how does it work?

Devices: They only looked at television, a magazine, or a newspaper, unexpectedly, and used personal devices to browse the internet for it. The most interested ones bought a book or borrowed it from a library.

Cybermaturity 1

This is the level that corresponds to Initiate, whether it is sheer ignorance, incompetence, or innocence. How did each of Omega's technology areas look during this stage?

People: They did not know who was interested. They decided there was no designated person in charge.

Systems: There were no systems related to digital presence, although it is unclear whether it was due to ignorance or a reluctance to "get into that technology stuff."

Processes: None.

Policies: There was no need for rules for something that did not exist.

Devices: The closest thing to digital presence was attending meetings with their fingers. No way to leave them at home.

I hope this overview of each technology area's journey through each stage of maturity is useful in illustrating the process.

It all starts with a small idea that matures over time. In the beginning, it may be important to try to imitate a more mature ministry or organization. That is okay. Over time, that initial imitation must give way to a unique identification where the organization's own distinct characteristics are discovered and expressed in a unique way.

Reverse engineering is appropriate for reducing the complexity of an object or system, discovering its flaws, and facilitating reuse.

This is the process of reverse engineering, which, even if it has not been formally defined in your organization, is at the essence of every new project.

I want to conclude with a few words about the importance of adjusting the maturity roadmap to your specific condition. This is the topic of the next and final chapter.

CUSTOM ROADMAP

What I am saying is that as long as an heir is underage, he is no different from a slave, although he owns the whole estate.[34]

– SAINT PAUL, APOSTLE.

There are two kinds of people in the world: the oriented and the disoriented. Spatial orientation, the ability to navigate in physical and geographical space, recognizing shapes, colors, perspective, and moving in three dimensions, is classified as one of the seven intelligences: Spatial intelligence. Well, my wife and I belong to each group separately.

How many people give glory to God for GPS? Before this useful technology entered our lives, my wife would easily get lost. I would make her a detailed map of the route and write down every right turn and left turn in detail. In my instructions, I included landmarks that were unmistakable to me, like a particular building or a billboard.

Despite my efforts to prepare her, she would still get lost. Then she would call me to ask for directions, in a conversation that went something like this:

"I'm lost. Can you help me?"

"Where are you?"

"I don't know. Let me see... —driving time passes by— Washington Boulevard."

"In which direction?"

"How would I know if I'm lost?"

"Well, then tell me which street you are crossing."

"This is —more driving time passes by— 5th Avenue. But don't get mad..."

"I'm not getting mad, I'm just taking the time to write but…"

"Now I'm on 6th. I made a right turn..."

"No! Go back to Washington Boulevard and make a left turn."

"I can't talk now, bye!"

I would be left hanging, not knowing where she would go until she called me again, saying, "I'm lost," and the cycle repeated. To be honest, it was frustrating. I could hardly ever give her instructions over the phone. I am very visual, and she is very verbal. Eventually, she decided not to call me anymore for directions and started calling a mutual friend, who was undoubtedly more patient than me.

It is not a gender issue. I know a couple where he is always lost, and she is the one who gives him directions to places.

I Am Lost

It is not something that only happens to individuals. A Christian organization or institution can find itself "lost" in its own circumstances when it comes to ministry technologies. You may have a user manual, detailed instructions on how to apply a particular technology, and have learned to turn the knobs of your devices to the right or left, and yet you may not find your place nor your destination.

The maturity map in this sense can be as overwhelming as the Thomas Guide. The Thomas Guide is a thick spiral-bound book with a detailed map of the entire city. I used to carry it in the car wherever I went. All the geographical information I needed was there in great detail, but if at the precise moment when I consulted it I did not know where I was, or in other words, my location on the map, I was definitely lost.

You may have a user manual, detailed instructions on how to apply a particular technology, and have learned how to turn the knobs of your devices to the right or left, and yet you may not find your place nor your destination.

The following five factors can help you determine if your organization is lost in technology matters:

1. *You are more focused on the technique than on the results.* It is like driving a car purely for the adrenaline rush and burning fuel. There is nothing wrong with becoming a master of the technique, but technology can be so enticing that it can trap you and make you lose sight of why you wanted to use it in

the first place. This happens when technology becomes the goal and displaces the ministry.

2. *You are going around in circles without progress.* You have seen the same sign several times and finally realize that the place looks familiar. Although maturity is a cyclical process, it is not a closed circle but a spiral; each turn should move you to a new level of progress. The only way to know if you are making progress is through constant evaluation.

3. *You do not know which stage you are in.* You need a minimum of two coordinates to establish your location. This is the most common problem I encounter in Christian organizations and the one that has driven me to write this book. You need to run the assessment from Chapter 8 at least twice to establish your coordinates. If you do not know where you are, it's very difficult, if not impossible, to give you directions that will lead you to your destination.

4. *You do not know what level of maturity you want to achieve.* So, it is not about the route, but the destination. It is okay if your institution does not want to become innovative in a certain technology, but what is not okay is if you come to this conclusion by chance or drifting. Your maturity destination needs to be planned and controlled, entrusted to God's hands rather than left to chance.

5. *You want to travel many paths at the same time.* This can be observed when you have many programs, but you do not delve deep into any of them. You may be navigating in shallow waters. There are many possible paths that can lead you to your destination, but you cannot take them all at the same time. You must decide which route you will take and what sort of things you will try.

Calling a Friend

There are three critical issues when consulting a map (which brings us back to the definition of cybermaturity). First, locating yourself on it; second, knowing which direction you are heading; and third, regularly evaluating the journey. You need to know these three things before you can correct your course.

If your organization is experiencing spatial blockage, it is likely that you have not been able to locate it on the map or know how to help it get back on track. You need to call a good friend, someone who knows your ministry well, who is more patient, and who is available when you feel lost.

Every Christian ministry that desires to be effective in carrying out its ministerial functions, within its geographic and historical context, must also ask the question: "Do I have the necessary resources to get from here to there?"

Rarely is the answer to this question "Yes." Frequently, the answer is "No." But most of the time, the answer lies somewhere in between: "I have some resources, but I need to supplement them."

It is in this scenario that your ministry needs to consider the possibility of calling a friend or hiring a consultant. A consultant is someone who brings expert opinion and has experience working with many organizations in a variety of circumstances. A consultant has the advantage of looking at the situation from the outside to offer a more panoramic understanding and develop a detailed roadmap for you.

I do not yet know the reason, but I often find that Christian ministries, especially smaller ministries, resist the idea of asking for help and being guided by experienced individuals. They struggle with the idea of inviting someone with expertise to assist them and find it difficult to justify the expense, without realizing that it is not an expense but a medium and long-term investment.

Hiring consultants is common in businesses. I can testify to this based on my over two decades of experience in the field of computer science. I have had the privilege of serving as a consultant for many higher education and professional institutions, and I have had the pleasure of instructing hundreds of individuals in the use of various technologies. At the same time, I have found myself in the position of complementing the expertise and work of the technology teams I have led with consultants and advisors.

> *A consultant has the advantage of looking at the situation from the outside to offer a more panoramic understanding.*

The Bible is filled with examples in which leaders have sought advice from individuals with more experience or knowledge. The first biblical manual of consultancy can be found in Exodus 18:13-27.

Consultancy Biblical Manual

Jethro, Moses' father-in-law, had come to visit and bring Moses' wife and children after Moses had liberated the Israelites from the rule of Pharaoh. The story narrates how Jethro observed everything Moses did throughout the day, noticing that he spent the entire day, from morning till evening, judging the people's affairs.

Jethro approached Moses with some questions and after listening carefully, he said, "What you are doing is not good" (Exodus 18:17). It is not good for a Christian ministry or organization to try to handle all the technological work by themselves, without being open to the possibility of seeking good advice or having the most suitable people to do the job.

Proverbs 11:14 establishes an inseparable relationship between the leadership of the ministry and the need for counselors: "For lack of guidance a nation falls, but victory is won through many advisers."[35]

Proverbs 15:22 affirms along the same lines the directly proportional relationship between plans and good advice: "Plans fail for lack of counsel, but with many advisers they succeed."

And Proverbs 24:6 confirms that a good strategy, one that leads any organization to overwhelming victory, is produced and directed through the multitude of counselors, saying: "Surely you need guidance to wage war, and victory is won through many advisers."

Why? What is it that makes many counselors refine your direction, execute your plans, and help you achieve your victories? What are the faculties that facilitate maturity? The answer is found, at least partially, in the person of Jethro, in what I like to call the biblical manual of consultancy. These are the observable characteristics of Jethro that every good ministerial advisor should have.

Do not hesitate to consider the idea, even during the first steps of cybermaturity. A good advisor could guide you with his or her expertise and save you many hours and headaches early in the process.

A Good Advisor Knows Your Successes

Jethro visited Moses after hearing about all that God had done through him (Exodus 18:1). A good advisor has done prior research that allows them to understand your strengths and the things that have led you to the place of success where you currently are. Sometimes we think we need counseling when things are already going wrong or getting out of hand, such as when a couple is on the verge of divorce, or a company is on the brink of bankruptcy. The best time to hire a good advisor is now, while you are enjoying success.

A Good Advisor Rejoices in Your Achievements

Jethro was glad to hear about his son-in-law's and the people's success in their confrontation with the Egyptians (Exodus 18:9). A good advisor reminds you of your past successes to build your confidence and place you in the right mental and emotional framework to reach your future goals. By doing this, they refine your strengths and help you make the most of them.

A Good Advisor is also a Good Priest

I do not mean this in the formal sense of some Christian traditions. I mean a priest in the sense of 1 Peter 2:9. Jethro was already a priest, and he recognized the Lord God working among the people through the hand of Moses (Exodus 18:10-12). A good priest/advisor praises God and affirms that God is with you. A good advisor/priest understands your higher calling, your service, and the impact it brings to your community. They know that God is guiding you in the process and they are willing to walk alongside you.

I have nothing against non-Christian advisors, but at some point, they may lose sight of your main purposes and your reason for being. The priest knows the right place of technology in ministry.

A Good Advisor is a Keen Observer

This is the advantage of having someone who sees things from the outside. Someone who is not immersed in the details or involved in daily procedures can always see things that those who deal with the daily work cannot see. Jethro saw (Exodus 18:14) –how powerful! What Moses, Aaron, Miriam, and the other two million Israelites could not see, the newly arrived Jethro did.

A good advisor takes the time to observe, and through observing, they see. By looking at what is right in front of their eyes, they can develop a vision of the future for you.

A Good Advisor Asks the Right Questions

There are difficult and confrontational questions that many are afraid to ask. It is possible that some people close to you may have already noticed the problem and thought about asking, but we will never know if they truly saw it because they never asked. The advisor has nothing to lose. They can ask what others would not, even if it may upset you. They can initiate those crucial conversations.

Jethro exclaimed with astonishment and reproach. One can sense his concern for Moses and the people: "What is this you are doing for the people? Why do you alone sit as judge, while all these people stand around you from morning till evening?" (Exodus 18:14).

> *The best time to hire a good consultant is now, while you are enjoying success.*

A Good Advisor Knows How to Listen

Has someone asked you a question and gotten distracted by something else when you try to respond? In those cases, you ask yourself, "Why did they ask if they do not want to know?" Jethro actively listened to Moses' response (Exodus 18:15-16). Want to evaluate a consultant to see if they are good? Observe how they listen. A good advisor wants to know... without rushing. Their questions are not rhetorical. They seek good actionable answers.

A Good Advisor is Not Afraid to Point Out the Problem

I had a boss with strong leadership skills. The first meeting I had with her after being hired as an employee taught me a valuable lesson that I still cherish to this day. She said, "Bad news does not get better with time." It is so true. An illness does not get better if you postpone addressing it. Divisions in the church do not dissolve if you let time pass. Every problem needs intervention. A good advisor will always pinpoint the issue, just as Moses' father-in-law did (Exodus 18:17). They eliminate the fear of criticism and the possibility of personal ego wounds. They acknowledge your problems and move you toward maturity.

A Good Advisor Understands the Magnitude of the Work

The normal tendency of any new project is to underestimate the effort required. In technical project management, I have practiced something that has never let me down. I listen to workers estimate the time it would take to complete a task, and then I double it to calculate my own estimates. If they say one day, I estimate two, if they say two weeks, I estimate a month. It always works. The work is bigger than we initially thought. A good advisor knows this. Do not be surprised if they use the same technique (Exodus 18:18).

A Good Advisor Proposes Accurate Solutions

This is what happened with Jethro. He divided Moses' giant workload into smaller pieces. He suggested a knowledge transfer system (Exodus 18:20) and personnel selection (Exodus 18:21). He classified the work between simple and more difficult cases and strengthened leadership (Exodus 18:22). A good advisor could relate their experience to your specific context and knows how to delegate work to solve problems. They would even guide you to delegate work even though you have not done it before.

A Good Advisor Recognizes When Their Work is Done

Every advisor makes a living from their consulting practice. The best thing for their own business is to extend their contract to earn more. But a good advisor is not opportunistic or exploitative. They do not waste your resources and they know when to stop; you do not need to send them home, they leave on their own accord. Jethro returned to his own country (Exodus 18:27) What a conscious father-in-law! If only they were all like that... the advisors, I mean.

Aubrey Malphurs, in his book *A New Kind of Church: Understanding Models of Ministry in the 21st Century*, provides seven reasons why your ministry can hire ministry consultants:[36]

1. Ask yourself if you can afford not to hire an advisor. Sometimes the costs of not doing so outweigh what you can pay someone else.
2. An advisor will have an influence on your future income. There is a direct relationship between ministerial maturity and financial improvements.
3. It is possible that someone is willing to donate the money to hire an advisor if the benefit is tangible.
4. Remember that you get the level of service you are willing to pay for. Cheap services can end up being costly. You can't put your ministry in the hands of a mediocre advisor.
5. If your ministry is small, you can join efforts with other small ministries to share expenses.
6. Evaluate the need to grow and complement your resources above the costs of hiring someone.
7. A good ministerial advisor will always be more efficient and help your ministry achieve better results.

The time has passed when it the pastor or leader was expected to know everything about everything and did everything. Our time demands the wisdom of many counselors.

I hope that the brief analysis presented here helps you have a clear criterion during your hiring process and that when you complement your work team by calling a friend, you can achieve the level of maturity that your technology initiatives require.

Before choosing an advisor, you need to subject them to a series of interviews, first with you, and then with the people who will be closely collaborating with them. Remember that people can learn technical skills, but having the right heart is the result of a maturity process that could take many years.

Appendix E provides an assessment table for advisors that you can use during your hiring process.

Custom Maturity Roadmap™

When you think about the maturity roadmap presented in this book and how you could implement a technology maturity initiative, some questions come to mind that need answers:

How much do we need to mature? What level do we need to reach considering our goals? What kind of growth do we want to see? How do we balance numerical growth with the maturity of ministerial technologies? How do we reach that desired level of maturity? And of course, as discussed in the previous section, do we have the necessary resources for the journey we need to undertake?

The answer will depend on the specific details of your life, your leaders' lives, and the life of your ministry. While it is essential to have a maturity model, that model must be grounded in concrete ways to accurately represent the realities of your ministry.

This is what I have called the *Custom Maturity Roadmap*™. It is the combination of the maturity model I have just presented and the specific way that model will take shape in your personal life, the lives of your leaders, the life of the ministry you serve, and the life of how the ministry utilizes technology.

The *Custom Maturity Roadmap*™ is a process of technical advisory that begins with an assessment of the current state of the use and development of ministerial technologies in your organization or church and culminates in a detailed document of your opportunities for maturity.

The maturity roadmap I have presented in this book is a general map and, if you will, universal. You can apply the roadmap to personal growth in areas such as character, spirituality, or interpersonal relationships. You can also apply it to how each individual uses technology in terms of productivity and organization. I have given it a corporate focus, applying it to every Christian organization.

It is possible to use the *Custom Maturity Roadmap*™ to mature specific technologies, just as I did with the example of Omega's digital presence in the previous chapter, and other technologies such as administrative, production, or media technologies.

People can learn technical skills, but having the right heart is the result of a maturity process that could take many years.

It is also possible, and recommended, to go through a process of strategic technology planning, applying the *Custom Maturity Roadmap*™ to ministerial technologies, considering all the technology that is currently in use or in the plans of your ministry. To that end, I suggest combining the *Custom Maturity Roadmap*™ with the integrated model of ministerial technologies that I have called the *Ministerial Technology Matrix*®. You can find more information about the matrix as a strategic and organizational concept on my website.

And with that, we come to the end of this chapter and our discussion. I encourage you to use the additional resources in the following appendices and share your feedback and progress with me. The path to maturity may be steep or challenging. You must embark on it with humility and meekness, and hopefully with the intentionality it requires. I guarantee you that it is worth it.

So, let us mature!

APPENDIXES

APPENDIX A: ATTITUDES SCALE

Scale: Attitudes towards technology maturity

Name ____________________ Age: ____ Gender: M F

This scale evaluates attitudes towards technological maturity in ministry. There is no right or wrong answer. What matters is your perspective. Read each statement carefully and indicate how much you agree with each one. Mark with an 'x' or circle 'o' the value that you consider most accurate for each attitude, ranging from -5 for immature attitudes to 5 for growth alternatives. We appreciate your patience and accuracy.

Name of the team member:											
Fear										Love	
-5	-4	-3	-2	-1	0	1	2	3	4	5	
Resistance										Meekness	
-5	-4	-3	-2	-1	0	1	2	3	4	5	
Conformity										Entrepreneurship	
-5	-4	-3	-2	-1	0	1	2	3	4	5	
Blindness										Vision	
-5	-4	-3	-2	-1	0	1	2	3	4	5	
Comfort										Innovation	
-5	-4	-3	-2	-1	0	1	2	3	4	5	
Sterility										Fruitfulness	
-5	-4	-3	-2	-1	0	1	2	3	4	5	
Grand total											

Appendix A: Attitudes Scale

Scale Totals

Name	Fear	Resistance	Conformity	Blindness	Comfort	Sterility	Grand Total
Total							
Averages (Total/)							

Appendix A: Attitudes Scale Totals by Team

If you want to obtain an electronic copy of the *Attitude Scale* and all other assessment tools in this book for printing, please visit: www.cyberministry.info.

APPENDIX B: QUESTIONNAIRE

Questionnaire: Principles Lanes

Name ________________________ Age: ____ Gender: M F

This questionnaire evaluates how the five principles of technological maturity govern technology programs in ministry. There is no right or wrong answer. What matters is your point of view. Read each statement carefully and indicate your level of agreement. Mark with an 'x' None - if you disagree; Some - if you somewhat agree; Moderate - if you somewhat agree but have doubts; or Strong - if you strongly agree. We appreciate your patience and accuracy.

#	What is your level of agreement with these statements?	None	Low	Some	High
1	We have been able to discern our calling to ministry and are taking steps to excercise it.	☐	☐	☐	☐
2	If I think about the ability of technology to prolong life, I can cite some examples in my mind.	☐	☐	☐	☐
3	The vision of the church is written in detail and the leaders understand were we are heading as an organization.	☐	☐	☐	☐
4	I aspire with confidence for the church ministry to progress to the next level of technological maturity.	☐	☐	☐	☐
5	The church seeks to explore different alternatives to solve the technological problems it faces on a daily basis.	☐	☐	☐	☐
6	Technology helps us live in devotion to God, knowing that our primary goal is to worship Him.	☐	☐	☐	☐
7	Everything we do and everything we should do must be guided by excellence; excellence glorifies God and adds more honor to His name.	☐	☐	☐	☐
8	There is a direct relationship between the technology I use and the growth I see in the ministry; better technology leads to greater growth.	☐	☐	☐	☐
9	Every time there is a change, it is easy for me to adjust my behavior and routine to adapt.	☐	☐	☐	☐
10	We have defined a roadmap that everyone can follow to become a disciple of Jesus Christ, and we actively follow it.	☐	☐	☐	☐

Maturity Principles Lanes (continued...)

#	What is your level of agreement with these statements?	None	Low	Some	High
11	The use of technology is critical to revolutionize the practice of ministry in the church.	☐	☐	☐	☐
12	I value the church's ability to bring to life people by communicating the gospel of Jesus Christ.	☐	☐	☐	☐
13	The church finds it easy to replace the old version of a program when the new one is released.	☐	☐	☐	☐
14	Our team enthusiastically aligns our values with our vision, mission, and strategies, understanding the role of technology.	☐	☐	☐	☐
15	Technology when applied to ministry improves all functions of the church in its mission.	☐	☐	☐	☐
16	We have a clear and memorable mission statement that our members can easily recite.	☐	☐	☐	☐
17	We understand the concept of the fruit of the Spirit and live consciously to grow spiritually in order to demonstrate that fruit as a church.	☐	☐	☐	☐
18	The best way to alter the material world in which the church operates is by adopting and modifying the use of certain technologies.	☐	☐	☐	☐
19	We find that our calling takes priority over activities and many engagements. Technology helps us to maintain our priorities in check.	☐	☐	☐	☐
20	Growing in quality is very important to demonstrate that the Holy Spirit is at work. Technology helps us to do work of a higher quality.	☐	☐	☐	☐
21	I feel my team maintains a set of balanced attitudes towards the use of technology in ministry.	☐	☐	☐	☐
22	I easily support the adjustments that every change brings to the ministry environment and the way of doing ministry.	☐	☐	☐	☐
23	Changes are very exciting. I like it when our ministry runs an evaluation to introduce certain changes.	☐	☐	☐	☐
24	Growing in quantity is important to demonstrate that the Holy Spirit is at work. Technology helps to involve more people in ministry.	☐	☐	☐	☐
25	We have clear goals to make a possitive impact in the community, knowing that the church is the hope of society.	☐	☐	☐	☐

If you would like to obtain an electronic copy of the *Principles Lanes* questionnaire and all other evaluation tools for printing, please visit the following website www.cyberministry.info.

Maturity Principles Lanes (continued…)

#	What is your level of agreement with these statements?	None	Low	Some	High
26	Technology provides us with capabilities to prosper and grow everything we are involved with.	☐	☐	☐	☐
27	We are disciplined individuals and strive to maintain balance in all areas of life and ministry with the help of technology.	☐	☐	☐	☐
28	It is my duty to participate with the church in the transformation of the world around us, starting with the immediate community.	☐	☐	☐	☐
29	We have included the role of technology in our strategic discussions during our ministry meetings.	☐	☐	☐	☐
30	Although I am aware of the dangers of technology, I prefer to think that technology primarily serves to protect life.	☐	☐	☐	☐
31	We promote higher purposes in life beyond mere survival, and technology helps us communicate our spirituality.	☐	☐	☐	☐
32	I consider myself as a vital agent, with the capacity to give promote, protect, prolong, and enhance the lives of other people.	☐	☐	☐	☐
33	I always seek opportunities to learn new technologies that facilitate ministerial work.	☐	☐	☐	☐
34	The church lives more to be than to do, that is, to forge the character of Jesus Christ in the lives of believers and to set Him as the model to follow.	☐	☐	☐	☐
35	Serving others is of utmost importance in the church and is an extension of God's mercy towards others.	☐	☐	☐	☐
36	I can use certain technologies in ministry to promote improvements in the lives of others.	☐	☐	☐	☐

Appendix B: Questionnaire: Principles Lanes

Questionnaire Answer Key

Use the key for the questions below to evaluate the responses for each of the five principles. For example, questions 4, 5, 8, 13, 21, and 33 from the previous questionnaire all correspond to the principle of substitution. Questions 3, 10, 14, 16, 25, and 29 can be considered control questions, but they are related to strategic use of technology in ministry matters.

Principle	Questions
Substitution	4, 5, 8, 13, 21, 33
Transformation	9, 11, 18, 22, 23, 28
Revitalization	2, 12, 15, 30,32, 36
Reproducibility	7, 17, 20, 24, 26, 34
Spirituality	1, 6, 19, 27, 31, 35
Strategy	3, 10, 14, 16, 25, 29

Appendix B: Questionnaire Key

Questionnaire Totals

Name	Substitution	Transformation	Revitalization	Reproducibility	Spirituality	Grand Total
Total						
Averages (Total/)						

Appendix B: Principles Lanes Totals by Team

APPENDIX C: TECHNOLOGY WHEEL

Assessment: Technology Wheel

Name ________________________ Age: ____ Gender: M F

This technology wheel evaluates the maturity of the five technology areas according to the seven steps of maturity. There is no right or wrong answer. What matters is your point of view. Mark with an 'x' or fill in the circle where each technology area intersects with the maturity level (1-7) you believe you have reached. We appreciate your patience and accuracy.

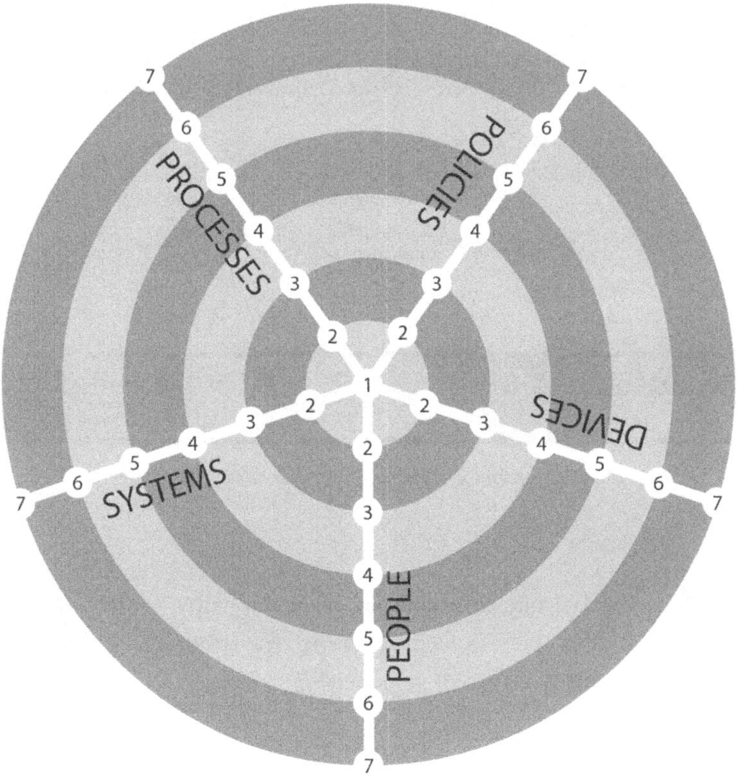

Appendix C: Technology Wheel

Technology Wheel Totals

Name	People	Systems	Processes	Policies	Devices	Grand Total
Total						
Averages (Total/)						

Appendix C: Technology Wheel Totals by Team

If you would like to obtain an electronic copy of the *Technology Wheel* and all other evaluation tools for printing, please visit www.cyberministry.info.

APPENDIX D: DISCUSSION QUESTIONS

This is not an exam; it is just a set of questions to facilitate group discussion. I have divided the questions into chapters to facilitate navigation and focus, but you do not need to use them in that way. You do not need to answer all the questions, but I recommend that you respond to at least two or three from each section during your planning and evaluation sessions.

Chapter 1: Maturity

1. What does maturity consist of?
2. According to Genesis 1:28, what is the first commandment, and how does it relate to maturity?
3. What is the relationship between being mature and being spiritual?
4. Is maturity important, necessary, or both? Why?
5. How does the concept of maturity relate to ministerial technologies?
6. Can you identify any areas of technology use in your ministry that need to go through a maturity process? Which ones and why?

Chapter 2: Immaturity

1. What is your perspective on immaturity? Is it bad? Is it good?

2. Is immaturity something we should strive to eradicate?
3. Can you identify areas of technological immaturity in your ministry?
4. What are the growth alternatives?
5. What are some attitudes of immaturity that you need to overcome in your ministry?
6. Why can you apply the concept of maturity, which is related to living beings, to ministerial technologies?
7. Read Matthew 11:29, what is the role of humility and meekness in the process of maturity?
8. Read 1 Corinthians 13:11-12, what is the relationship of this passage to personal maturity and how can you apply it to technological maturity?

Chapter 3: Technology

1. Why is it difficult for us to identify certain technologies as such due to our familiarity with them?
2. What effect does taking certain technology for granted have on our relationship with it?
3. In your own words, what is technology?
4. How is your concept of technology similar or different to the concept presented in the book?
5. What significance does the integrated ecosystem of technology have for your ministry?
6. What significance does have the fact that technology is a cultural network for your ministry? What are the implications for your cultural context?
7. Choose a technology in use in your ministry and identify each of the five areas in detail.
8. What specific purpose does technology serve in your ministry? Have you articulated such a purpose? Does it include the three elements of strategy, tactics, and intelligence?
9. What do you think about the statement: "It is impossible to do ministry without technology"? Why?

Chapter 4: Maturity Models

1. Can you mention three maturity models that you can observe in nature?
2. Why do you think technological maturity is often unknown in ministerial circles?
3. Can you mention three existing models of technological maturity?
4. To what or whom does the concept of technological maturity apply?
5. How can you apply the maturity model of your calling to your life? Provide concrete examples. How does this relate to the maturity of technology?
6. What are the principles that govern technological maturity? Which of them can you identify with the most? Why?
7. Can you provide some examples of how you find these principles at work in your current ministry?

Chapter 5: Maturity Roadmap

1. What is one of the primary functions of a leader? Why?
2. Looking at the seven steps of the technology maturity roadmap, with which one can you identify the most? Why?
3. In which of the seven steps do you think your current ministry is currently situated? Compare your answer with other members of your ministry team. Are they different or the same? Why?
4. What steps could move you to a higher level of technological maturity in your ministry context? Be as specific as possible.
5. What is the difference between ignorance, uncertainty, incompetence, and innocence?

6. What comment can you make regarding the phrase: "Intentions are not enough"?
7. What are some characteristics of the Ideation stage?
8. What role does the development of a personal identity play in technological maturity?
9. What is the most advanced stage of technological maturity and why? How does this stage relate to the definition of maturity and spirituality?
10. Read Ecclesiastes 1:9. Was Solomon right? What attitudes contrary to maturity can we discover in his statement?

Chapter 6: God's Maturity

1. John 2:1-11 narrates the first miracle of Jesus. What parallel exists between the maturity of wine and technological maturity?
2. Can we do theology considering the maturity of God? Why? Why not?
3. What does biblical perfection consist of? From this perspective, can you become perfect? Can a technology program become perfect?
4. Discuss the phrase: "God is the most mature being in the universe."
5. Read Daniel 7:9, 13, 22 and 1 Corinthians 3:1-3. What is the relationship between what the passage says and the description of God as the Ancient of Days? How does this relate to maturity? Can you apply these passages to the maturity of your technology programs?
6. Discuss the phrase: "God is the most fruitful being in the universe."
7. Read Hebrews 12:2 and answer, in what ways is Jesus Christ a model of maturity?
8. Read Galatians 5:22-23 and Ephesians 4:13. In what ways is the Holy Spirit a model of maturity?

9. How is the maturity roadmap applied to God? In which of the stages is He? Has He ever been in any of the other stages?
10. Read Isaiah 43:19 and Revelation 21:5. In what way does the prophet Isaiah differ from King Solomon in their appreciation of the new? According to these passages, why is God the most mature being in the universe? What do these passages tell us about the maturity of God?

Chapter 7: Cybermaturity

1. Have you experienced difficulties and setbacks in implementing your technology programs in ministry? List your setbacks and explain.
2. Based on the origin of the word "cybernetics" and the account in Acts 27, what is the relationship between leadership and maturity?
3. What do navigation and control instruments have to do with maturity?
4. Do you have navigation instruments in your ministry? List them.
5. Take a blank sheet of paper and divide it into three parts. Write the headings: Location, Direction, and Measurement. Now try to define these elements in your ministry.
6. What is technology? How does the concept of technology relate to technological maturity?
7. Which areas of technology in your ministry are the most deficient? Why?
8. What is cybermaturity and how can you apply it to your ministry?

Chapter 8: Assessing Maturity

1. Why is measurement important in technology maturity?

2. What type of evaluation do you regularly conduct in your ministry? If you do not have any, why not?
3. Name some reasons why evaluating your ministry as a team is more effective.
4. From the action list in the Teamwork section of **Chapter 8**, what will you implement today?
5. Write down the names and phone numbers of each member of your technology team along with their respective roles or responsibilities.
6. What can you learn from the attitude scale towards technology? Who are the optimistic members of your team? And the pessimistic ones?
7. What can you learn from the lanes of maturity principles in technology maturity?
8. What can you learn from the technology wheel about the state of technology in your ministry? How does this differ from your initial impressions gathered in question 7 from the previous section in **Chapter 7**?
9. What are your overall impressions of the technology maturity evaluation in your ministry?
10. With the evaluation in hand, go back to the maturity map in **Chapter 5**. The next level is your most immediate goal. What actions can you take today to move to the next level?

Chapter 9: Applying Maturity

1. What is reverse engineering and how can you apply it to your ministry?
2. What other ministries do you consider more mature than yours in certain areas of technology that can serve as a basis for your innovation process? Make a list of their names and their model technologies.
3. What elements specific to your identity, reality, and particular context do you need to incorporate into these models to recreate something new and unique?

4. Think about your own digital presence. Is this an area of ministry technology that you need to mature? How do you plan to do it?
5. How can you analyze the maturity of each technology area in each stage that you have 1. Reached and 2. Want to achieve?
6. If you want to visualize this, take a sheet of paper, and divide it into two columns with the headers: Now and Future, respectively. Then, write down a specific technology that you are using for ministry purposes or planning to use and break it down into its five components. Analyze the level of maturity, from 1-7, and determine the steps you can take to grow in maturity.

Chapter 10: Custom Roadmap

1. What are the five factors that can help you determine if your ministry is lost in technology matters? Which of these applies most to your reality?
2. Why can it be important for your ministry to hire an advisor?
3. Read Exodus 18. Jethro, Moses' father-in-law, was an excellent advisor. Why?
4. Of the nine characteristics observed in the Biblical Advisory Manual, which resonate most in your ministry and why?
5. What are the most common objections you have within leadership to not hiring an advisor? How can you address these objections?
6. Do you understand the need to create a personalized roadmap? Why or why not?

If you wish to obtain an electronic copy of the discussion questionnaire and all other evaluation tools for printing, visit www.cyberministry.info.

APPENDIX E: CONSULTANTS INTERVIEW

Use the following nine questions in this section to interview all your ministerial advisors in general, and your ministry technology advisors, before hiring them.

To conduct this interview, it is advisable to present each question in a specific scenario. Listen carefully for clues that will help you evaluate the interviewee using a scale of 0-3 for each area.

#	Name: ______________________	0	1	2	3
1	Has the consultant done enough research on you as to know your sucesses and strengths?	☐	☐	☐	☐
2	Is the consultant conscious of your accomplishments and refers to them with a positive attitude?	☐	☐	☐	☐
3	Does the consultant show good evidence that show he or she is a good priest, love God and others?	☐	☐	☐	☐
4	Is the consultant a good observer, meticulous, pacient, and pays attention to detail?	☐	☐	☐	☐
5	Does the consultant know to ask the important questions? Asks more often than responds?	☐	☐	☐	☐
6	Is the consultant an active listener, provides good feedback, and ensures he or she understood?	☐	☐	☐	☐
7	Is the consultant brutally honest when pointing to a mistake he or she made?	☐	☐	☐	☐
8	Does the consultant seek understanding of the work at hand or tries to minimize it to avoid alerting you?	☐	☐	☐	☐
9	Does the consultant offer you an exit strategy? Does he or she know when the work is complete?	☐	☐	☐	☐

Appendix E: Consultants Interview

If you would like to obtain an electronic copy of the *Consultants Interview* and all other evaluation tools for printing, please visit www.cyberministry.info.

ACKNOWLEDGMENTS

To God, because I owe everything to Him. He is the most mature Being, who instills in us the desire to mature and who will bring all our maturity endeavors to their fullest potential to bear fruit.

To my wife, Ivonne Cabral, my toughest yet selfless critic. Her advice always helps me think practically, from the reader's perspective. This work would not be possible without all the support she provides. To my daughters, Johanna Lugo, and Natasha Lugo, for allowing me to borrow their time and lending me their ears to clarify my ideas.

To my beta readers: my brother Randy Lugo, and my good friends Harold Caicedo, Jairo Ospina, Marcel Ponton, and Salvador Ramírez, for their support, loyalty, and dedication; their generosity fills me with hope. This is a better book thanks to their contributions and all their feedback.

To Sarai Lara, for her contribution in developing the Technology Wheel as an evaluation tool and for her input as a beta reader as well.

To the graphic artist, my compatriot Jesús Cordero, from *Anointing Productions*, who designed the book cover. Divine providence brought us together. To Carlos Callegari for the photo on the back cover, but more importantly, for his valuable advice throughout this adventure.

Finally, to Jon Teichrow and his wife Claudia Naciff, for the hours of solitude they offer me in their cabin to clear my mind and write its derivatives. The seeds of this book germinated there.

CONSULTING SERVICES

There are currently several recognized Christian consultants who assist churches and other organizations in various areas of ministry. However, the availability of Christian consultants trained as theologians and technologists at the same time is limited.

The need to address the technological demands of 21st-century organizations requires what I call "bridge people" who have a good understanding of the role of technology but can develop technology plans with theological sensitivity.

That is why I have developed a model of computer consulting services known as the *Ministry Technologies Matrix®* to assist you and your ministry in three fundamental areas:

- Coach: Associated with training staff in the technical tasks of using and applying various technologies, providing training for both technical and ministry personnel.

- Construct: Associated with establishing a strong digital presence through the implementation of websites, email, social media, and mobile applications.

- Consult: Associated with implementing the *Ministry Technologies Matrix®* within your ministry by creating a *Custom*

Maturity Roadmap™ that applies the concepts of this book to guide you towards fulfilling your mission in context.

In addition, I have created a maturity certification program that is based on a series of trainings, the evaluation tools described in this book, and your *Custom Maturity Roadmap*™. This program assigns a badge or seal that represents the level of maturity achieved by your organization or ministry, with various membership benefits at various levels.

I invite you to contact me to explore each of these three service areas I offer or the certification program. I can customize my work for short-term, medium-term, or long-term engagements, depending on your current needs.

You can use my online contact page for more information:

https://vladimirlugo.com/contact/

ABOUT THE SERIES

The topic of technology is extensive. That is why we need to explore the use of technology in church ministry from various angles.

Each volume in this series provides a theoretical framework, born from theological and managerial reflection on technology, while also offering practical advice based on technological and ministerial experience.

In other words, each book will show you the dock and how to swim to stay safe in these times of turbulent change.

Cyberministry 1: *Cyberministry: 5 Essential Functions of Technology in the Mission of the Church*

Cyberministry 2: *Cybermaturity 7 Steps to Unlock the Power of Technology for Fruitful Ministry*

Also available in Spanish as *Ciberministerio* and *Cibermadurez*.

Visit this website if you want to stay informed about this series and other books in the collection:

www.cyberministry.info

ABOUT THE AUTHOR

I am a theologian by vocation and a technologist by profession.

– VLADIMIR LUGO.

Vladimir Lugo is a ministerial technology advisor, IT professional, writer, speaker, and minister by calling, who assists Christian organizations in adopting online ministerial technologies to unleash the power of their mission.

As an IT professional, he has over twenty years of experience. He has developed a deep understanding of information systems and has helped many institutions align their use of technology with their respective missions and strategic goals.

As a minister by calling, he has served the Christian community for over thirty years in ministry, understanding his calling as an educator and mentor. He has early adopted computerized programs for Bible study, social media, blogs, and electronic applications for teaching and learning. He has assisted various ministries in implementing administrative, productivity, and online presence systems. He studied Chemical Engineering in his home country, Venezuela, and graduated with a Master of Divinity (M.Div.) from Fuller Theological Seminary in Pasadena, California, with a concentration in Intercultural Studies. He holds specialist certifications in various Oracle technologies and as an *Oracle Cloud* and *Amazon Web Services* Certified Solutions Architect.

He worked for 13 years at Loyola Marymount University, where he managed the software infrastructure for all administrative systems, which is a technical way of saying that if there was a database and a web server, he was responsible for it.

He worked for Ellucian, a company that provides software solutions for close to three thousand universities and colleges, migrating these systems to the public cloud before forming his own IT consulting company, Tepui Cloud, and content development company, Tepui Media, for which he currently works. He spends his time reading because he believes that readers deserve the best. You can follow his reading habits on the social reading platform *Goodreads*:

https://goodreads.com/vladimirlugomt

He dedicates time to writing about leadership and the use of technology in Christian ministry through his blog, www.vladimirlugo.com. There, you can share your email to receive his articles: Cyberminister today, where he explores the topic of ministerial technologies week after week.

Currently, he is preparing other books for publication. By sharing your email, you will also stay informed about his future projects and opportunities to hear him speak live.

Together with his wife of over twenty-five years, Ivonne, they are parents to two daughters, Johanna, and Natasha. They live in the Silicon Beach area of California. He enjoys running, cycling, and swimming and frequently participates in triathlons and marathons. He loves God, loves his women, loves the Church, and loves the adventure of life.

You can check his detailed work and service history on his LinkedIn profile:

https://linkedin.com/in/vladimirlugomt

He is active on social media with the unique handle *@vladimirlugomt*. Join the conversation on social media about the book using the hashtag #cybermaturity and about ministerial technologies using the hashtag #ministerialtechnologies.

If you would like to invite him to speak at your next event:

https://vladimirlugo.com/contract/conference/

Reach him out via electronic mail at: vladimir@vladimirlugo.com or by scanning the following QR code:

LIMITATION OF LIABILITY

The author authored this book to provide information. He has made every effort to ensure that the information is accurate and comprehensive, but there may be typographical or content errors.

The purpose of this book is to educate. The author and the publisher do not guarantee that the information contained in this book is complete and shall not be liable for any errors or omissions.

The author and the publisher limit their legal liability for damages caused or alleged to be caused, directly or indirectly, by the content of this book.

If you have any claims or legal questions, please contact the author at: legal@vladimirlugo.com.

BIBLIOGRAPHY

Books

Álvarez, Francisco J., *Interpretación del Modelo de Madurez de Capacidades CMM para Pequeñas Industrias de Software*, Universidad Autónoma de Aguascalientes (Aguascalientes, México: 2008).

Arndt, W., Danker, F. W., y Bauer, W., *A Greek-English lexicon of the New Testament and other early Christian literature*, 3rd ed., University of Chicago Press (Chicago, IL: 2000).

Bandler, Richard & John Grinder, *The Structure of Magic I: A Book About Language and Therapy, Science and Behavior* Books, Inc., (Palo Alto, CA: 1975).

Bergson, Henri, *Creative Evolution*, Arthur Mitchell, trans., Henry Holt and Company (New York, NY: 1911).

Buytendijk, Frank, *Performance Leadership: The Next Practices to Motivate Your People, Align Stakeholders, and Lead Your Industry*, McGraw Hill (San Francisco, CA: 2009).

Cameron, William B., *Informal Sociology: A Casual Introduction to Sociological Thinking*, Random House (New York, NY: 1963).

Carlson, Curtis R. y William W. Wilmot, *Innovation: The Five Disciplines for Creating What Customers Want*, Crown Business (New York, NY: 2006).

Davis, Don, Eugene Patronis y Pat Brown, *Sound System Engineering*, 4th edition, Focal Press (New York, NY: 2013).

Goins, Jeff, *The Art of Work: A Proven Path to Discovering What You Were Meant to Do*, Thomas Nelson (Nashville, TN: 2015).

Heidegger, Martin, "The Question Concerning Technology," en *Basic Writings: from Being and Time (1927) to The Task of Thinking (1964)*, ed. David F. Krell, Harper & Row Publishers (San Francisco, CA: 1977).

Jarred, Alfred, *Exploits & Opinions of Dr. Faustroll, Pataphysician: A Neo-Scientific Novel*, Exact Change (Boston, MA: 1996)

Keener, Craig S., *Acts: An Exegetical Commentary: Volume 4, 24:1-28:31*, Baker Academic (Grand Rapids, MI: 2015).

Kerzner, Harold, *Project Management: A Systems Approach to Planning, Scheduling, and Controlling*, Tenth Edition, John Wiley & Sons (New York, NY: 2009).

Lugo, Vladimir, *Cyberministry: 5 Essential Functions of Technology in the Mission of the Church*, Tepui Media (Sheridan, WY: 2023).

Malphurs, Aubrey, *Advanced Strategic Planning: A 21st Century Model for Church and Ministry Leaders*, Baker Books (Grand Rapids, MI: 2013).

__, *A New Kind of Church: Understanding Models of Ministry in the 21st Century*, Baker Books (Grand Rapids, MI: 2007).

Morgan, Tony, *Developing a Theology of Planning*, Tony Morgan Live, LLC (Dallas, GA: 2013).

Nolte, William L., *Did I Ever Tell You about the Whale? Or Measuring Technology Maturity*, Information Age Publishing, Inc. (Charlotte, NC: 2008).

Paulk, Mark C., Charles V. Weber, Bill Curtis, Mary B. Chrissis, eds., and others; *The Capability Maturity Model: Guidelines for Improving the Software Process*, Carnegie Mellon University, The Software Engineering Institute Series in Software Engineering, Addison Wesley Longman (Menlo Park, CA: 1994).

Rogers, Everett M., *Diffusion of Innovations*, Fifth Edition, Free Press (New York, NY: 2003).

Ruiz Bueno, Daniel, *Padres Apostólicos*, Biblioteca de Autores Cristianos, Edición Bilingüe Completa, La Editorial Católica (Madrid, España: 1950).

Strong, James, *The New Strong's Exhaustive Concordance of the Bible*, Thomas Nelson (Nashville, TN: 1984).

Taylor, Jerome (tr. ed.), *The Didascalicon of Hugh, of Saint-Victor: A Medieval Guide to the Arts*, Columbia University Press, Reprint edition (New York, NY: 1991),

The New Revised Standard Version with Apocrypha, Division of Christian Education of the National Council of the Churches of Christ in the U.S.A., (New York, NY: 1989).

Thomson, William, "Electrical Units of Measurement," in *Popular Lectures and Addresses Vol. 1: Constitution of Matter,* Nature Series, MacMillan & Co. (New York: NY, 1889).

Yonge, C. D. with Philo of Alexandria, *The works of Philo: complete and unabridged*, Peabody (Hendrickson, MA: 1995).

Websites

Bible Gateway, HarperCollins Christian Publishing, [https://www.biblegateway.com], accessed for the last time May 5, 2023.

Lugo, Vladimir, *3 Aspects to Ponder about the Challenge of Technical Maturity*, (Los Angeles, CA: 2015) [https://vladimirlugo.com/3-aspects-to-ponder-about-the-challenge-of-technical-maturity/], accessed for the last time May 5, 2023.

__, *5 Components Your Strategic Plan Cannot Leave Out* (Los Angeles, CA: 2015) [https://vladimirlugo.com/5-components-your-strategic-plan-cannot-leave-out/], accessed for the last time on May 5, 2023.

__, *5 Columns that Sustain the Use of Technology in the Church* (Los Angeles, CA: 2016) [https://vladimirlugo.com/5-columns-that-sustain-the-use-of-technology-in-the-church/], accessed for the last time on May 5, 2023.

__, *4 Elements your Digital Presence Cannot Live Without* (Los Angeles, CA: 2015) [https://vladimirlugo.com/4-elements-your-digital-presence-cannot-live-without/], accessed for the last time on May 5, 2023.

__, *Rodéate de muchos consejeros* (Los Angeles, CA: 2016) [https://oidopastoral.com/rodeate-de-muchos-consejeros/], accessed for the last time on May 5, 2023.

Merriam Webster Dictionary, 2011 [https://www. merriam-webster.com], accessed for the last time May 5, 2023.

ENDNOTES

All translations into English by Vladimir Lugo.

[1] Lugo, Vladimir, *5 Components Your Strategic Plan Cannot Leave Out* (Los Angeles, CA: 2015) [https://vladimirlugo.com/5-compo nents-your-strategic-plan-cannot-leave-out/], accessed for the last time on May 5, 2023.

[2] Lugo, Vladimir, *5 Columns that Sustain the Use of Technology in the Church* (Los Angeles, CA: 2016) [https://vladimirlugo.com/5-co lumns-that-sustain-the-use-of-technology-in-the-church/], accessed for the last time on May 5, 2023.

[3] I follow a detailed discussion on how I arrive at this concept of technology in my book *Cyberministry: 5 Essential Functions of Technology in the Mission of the Church*, Tepui Media, LLC (Sheridan, WY: 2023), pp. 45-73. I also explore it in other books of the series by the same name.

[4] Merriam Webster Dictionary, *policy*, 2011 [https://www. merriam-webster.com], accessed for the last time May 5, 2023.

[5] Heidegger, Martin, "The Question Concerning Technology," in *Basic Writings: from Being and Time (1927) to The Task of Thinking (1964)*, ed. David F. Krell, Harper & Row Publishers (San Francisco, CA: 1977), p. 288.

[6] Lugo, *Cyberministry*, Óp. Cit., pp. 75-113.

[7] Davis, Don, Eugene Patronis and Pat Brown, *Sound System Engineering*, 4th edition, Focal Press (New York, NY: 2013), p.180.

[8] Nolte, William L., *Did I Ever Tell You about the Whale? Or Measuring Technology Maturity*, Information Age Publishing, Inc. (Charlotte, NC: 2008), pp. 20-25. The title of the book points to the fact that technological growth is slow in the beginning stages, gets accelerated in a second stage, finds a stable state in the middle, and then abruptly disappears, following a curve in the shape of whale.

[9] Paulk, Mark C., Charles V. Weber, Bill Curtis, Mary B. Chrissis, eds., and others; *The Capability Maturity Model: Guidelines for Improving the Software Process*, Carnegie Mellon University, The Software Engineering Institute Series in Software Engineering, Addison Wesley Longman (Menlo Park, CA: 1994), p. 9.

[10] Rogers, Everett M., *Diffusion of Innovations*, Fifth Edition, Free Press (New York, NY: 2003), p. 138.

[11] Taylor, Jerome (tr. ed.), *The Didascalicon of Hugh of Saint Victor: A Medieval Guide to the Arts*, Columbia University Press; Reprint edition (New York, NY: 1991), p. 56. It literally reads: "Ingenious want hath mothered all the arts." However, Hugh refers to what was known in the Middle Ages as the practical arts, which are equivalent to our modern-day concept of technology.

[12] Kerzner, Harold, *Project Management: A Systems Approach to Planning, Scheduling, and Controlling*, Tenth Edition, John Wiley & Sons (New York, NY: 2009), p. 58.

[13] *Ibid.,* pp. 928-929.

[14] Goins, Jeff, *The Art of Work: A Proven Path to Discovering What You Were Meant to Do*, Thomas Nelson (Nashville, TN: 2015), p. XXV.

[15] *Ibid.*, p. 195.

[16] Nolte, *Óp. Cit.*, pp. 119-120.

[17] Bandler, Richard & John Grinder, *The Structure of Magic I: A Book About Language and Therapy*, Science and Behavior Books, Inc., (Palo Alto, CA: 1975), p. 7.

[18] Carlson, Curtis R. y William W. Wilmot, *Innovation: The Five Disciplines for Creating What Customers Want*, Crown Business (New York, NY: 2006), p. 40.

[19] Digital refers to the use of electronic devices that process information based on binary digits (1/0), which correspond to the physical world in terms of the on (1) and off (0) states of an electric current. Digital culture establishes interpersonal relationships in terms of these electronic means. Analog or analogue culture is less mediated. It prefers face-to-face and traditional communication.

[20] Morgan, Tony, *Developing a Theology of Planning*, Tony Morgan Live, LLC (Dallas, GA: 2013), Kindle loc. 228.

[21] Arndt, W., Danker, F. W., & Bauer, W., *A Greek-English lexicon of the New Testament and other early Christian literature*, 3rd ed., University of Chicago Press (Chicago, IL: 2000), pp. 573–574. *Kybernétes*. For a deeper discussion, you can review Keener, Craig S., *Acts: An Exegetical Commentary: Volume 4, 24:1-28:31*, Baker Academic (Grand Rapids, MI: 2015), Acts 27:11.

[22] Arndt, et. Al., *Ibid.*

[23] Yonge, C. D. with Philo of Alexandria, *The works of Philo: complete and unabridged*, Peabody (Hendrickson, MA: 1995), p. 259.

[24] Ruiz Bueno, Daniel, ed., *Padres Apostólicos*, Biblioteca de Autores Cristianos, Edición Bilingüe Completa, La Editorial Católica (Madrid, España: 1950), p. 686.

[25] *The New Revised Standard Version with Apocrypha*, Division of Christian Education of the National Council of the Churches of Christ in the U.S.A., (New York, NY: 1989), 4 Maccabees 7:1.

[26] Merriam Webster Dictionary, *cybernetics,* 2011 [https://www.merriam-webster.com], accessed for the last time May 5, 2023.

[27] Buytendijk, Frank, *Performance Leadership: The Next Practices to Motivate Your People, Align Stakeholders, and Lead Your Industry*, McGraw Hill (San Francisco, CA: 2009), p. 34.

[28] In Acts 6:1-2, Luke narrates how the apostles had to select individuals to take care of the service to others so that they could devote themselves to the ministry of the word. Exodus 18, a passage that I study in detail in Chapter 10, recounts a similar experience when Moses had to delegate ministerial work to other counselors to attend to the more challenging cases.

[29] Jarred, Alfred, *Exploits & Opinions of Dr. Faustroll, Pataphysician: A Neo-Scientific Novel*, Exact Change (Boston, MA: 1996), p. 103.

[30] Thomson, William, "Electrical Units of Measurement," in *Popular Lectures and Addresses Vol. 1: Constitution of Matter*, Nature Series, MacMillan, and Co. (New York: NY, 1889) p. 81.

[31] Cameron, William B., *Informal Sociology: A Casual Introduction to Sociological Thinking*, Random House (New York, NY: 1963), p. 13.

[32] Bergson, Henri, *Creative Evolution*, Arthur Mitchell, trans., Henry Holt, and Company (New York, NY: 1911), p. 7.

[33] Lugo, Vladimir, *4 elements your digital presence cannot live without* (Los Angeles, CA: 2015) [https://vladimirlugo.com/4-elements-your-digital-presence-cannot-live-without/], accessed for the last time on May 5, 2023.

[34] Galatians 4:1.

[35] Lugo, Vladimir, *Rodéate de muchos consejeros* (Los Angeles, CA: 2016) [https://oidopastoral.com/rodeate-de-muchos-consejeros/], accessed for the last time on May 5, 2023.

[36] Malphurs, Aubrey, *A New Kind of Church: Understanding Models of Ministry in the 21st Century*, Baker Books (Grand Rapids, MI: 2007), pp. 188-189.

This book has been published by Tepui Media.

Tepui Media is a company committed to strengthening the initiatives of independent authors and amplifying their communication impact. Tepui Media has a team of highly qualified professionals in multimedia communications, including specialists in digital presence and marketing, audiovisual production, content strategy, and software development.

If you need to publish your book and create a strong platform for promoting and disseminating your message, talk to the experts:

https://tepui.media

info@tepui.media

Vladimir Lugo

Made in the USA
Monee, IL
04 November 2026

67384103R00138